Resource and Environmental Sciences Series

General Editors:
Sir Alan Cottrell, FRS
Professor T. R. E. Southwood, FRS

Also in this series:

Environmental Toxicology – John J. Duffus
Minerals from the Marine Environment – Sir Peter Kent

Ecology for Environmental Science:

Biosphere, Ecosystems and Man – J. M. Anderson
Food, Energy and Society – David and Marcia Pimentel
Environmental Biology – E. J. W. Barrington
Environmental Economics – Sir Alan Cottrell

Environmental Chemistry

R. W. Raiswell, P. Brimblecombe, D. L. Dent and P. S. Liss

School of Environmental Sciences, University of East Anglia

Edward Arnold
A division of Hodder & Stoughton
LONDON MELBOURNE AUCKLAND

© 1980 R.W. Raiswell, P. Brimblecombe, D.L. Dent and P.S. Liss

First published in Great Britain 1980
Reprinted 1984, 1988, 1991

British Library Cataloguing in Publication Data

Environmental chemistry. — (Resource and
  environmental sciences series).
  1. Environmental chemistry
  I. Liss, P S   II. Series
  574.5   QD31.2

  ISBN: 0 7131 2790 2

Typeset in India by The Macmillan Co. of India Ltd, Bangalore — 560001.
Printed and bound in Great Britain for Edward Arnold, a division of
Hodder and Stoughton Limited, Mill Road, Dunton Green, Sevenoaks,
Kent TN13 2YA by Athenaeum Press Ltd, Newcastle upon Tyne.

Preface

The growth of man's agricultural, industrial and urban activities has caused the appearance of many environmental problems which share common, disturbing characteristics. Such problems often appear unexpectedly and, as in the case of photochemical smog or the Freon-ozone interaction, frequently result from the apparently innocuous actions of ordinary people. In attempting to alleviate these problems it is vital to understand how the chemical environment behaves in its unperturbed state. Studies of natural environmental processes can provide the necessary base-line data to assess the impact of pollutants, whether these are totally manmade chemicals (e.g. the Freons) or natural substances whose concentrations are being increased (e.g. CO_2).

However, although the rapid development of modern analytical techniques has stimulated an exponential increase in the number and sophistication of environmental studies, there has not been a commensurate growth in understanding the behaviour of natural and manmade chemicals in the environment. This is partly due to the unreliability of much of the data, for example difficulties often occur in maintaining the integrity of samples during collection and storage, but also arises because ease of measurement has encouraged data acquisition at the expense of understanding. The basic aim of the discipline of environmental chemistry is to understand, and therefore systematise, environmental data in terms of chemical principles. By establishing how the existing and well-documented body of pure chemical knowledge can be applied to processes occurring in nature, it becomes possible to understand the behaviour of manmade substances in the environment.

This does not mean that environmental processes can be described by chemical principles alone. It is often necessary to utilise basic knowledge from other disciplines, such as biology, physics and geology. Environmental chemistry provides a way of looking at the environment from a chemical point of view, just as other books in this series present a perspective of the environment from other standpoints, e.g. environmental biology, ecology and economics. None of these presents the whole story, only a particular way of looking at the very complex environment in which we live.

This book has its origins in a short lecture and practical course

given by the authors to first year students in the School of
Environmental Sciences, University of East Anglia. The book is
similar in scope, but rather more detailed than the lecture course.
Here we have adopted the earth-air-water factory as an analogue to
illustrate the way in which chemical principles operate in the
environment. In the lecture course the framework is the hydrological
cycle and the chemical processes which occur as water, with its
dissolved and particulate load, moves from the atmosphere onto the
land surface, into rivers, lakes and oceans and is eventually
incorporated into marine sediments. However since the hydrological
cycle represents a useful and logical sequence for the presentation of
material, it is maintained in the ordering of chapters in this book.
For readers who do not have an extensive background in
chemistry, a glossary of terms is provided. These terms are indicated
by '†' on their first appearance in the text. This is only a very
cursory attempt at explaining the necessary chemical principles and
readers who find this insufficient are advised to consult a good
elementary chemistry text (a list of possibilities is given in the
Bibliography). Although aimed at first year students studying
environmental sciences, chemistry, geology, biology, or other science
subjects, this book should also appeal to sixth formers studying
chemistry or other sciences to 'A' level, as well as to anyone with (or
willing to acquire) a basic knowledge of chemistry and interested in
how the natural environment operates as a chemical system. Readers
whose appetite for the subject has been whetted by reading this
book will find quite extensive references to further reading in the
Bibliography.

Norwich   R. W. Raiswell
1979   P. Brimblecombe
                                                 D. L. Dent
                                                 P. S. Liss

Contents

1 Introduction

The growth and awareness of environmental chemistry parallels the emergence of chemistry as a rational discipline in the late eighteenth and nineteenth centuries. At first chemistry passed through a descriptive and experimental period in which important discoveries were made in deducing the properties and reactivity of the elements and their compounds and in their applications to industries such as smelting, metal-refining, glassware and pottery. Data collection and application were, however, insufficient to provide the basis of a modern science, in which results can be organised into a coherent structure with predictive properties. The publication of Mendeleev's Periodic Table in 1869 initiated the development of chemistry as a modern science, now concerned predominantly with understanding the principles which formalise the structure and behaviour of matter.

The evolution of environmental chemistry has so far followed a similar path, although progress has been slow. A century ago the ignorance of mining engineers left an obvious legacy in the spoil-heaps of poor-grade ore minerals which were exposed to surface weathering action. The solid residues of the weathered ore minerals stain the surrounding rocks in multiple hues of blue, green and ochre; and the soluble metallic components drain away, sometimes with disastrous effects on the surrounding aquatic ecosystem. It is now well-known that these toxic, metal-rich waters result from the formation of soluble metal salts by atmospheric oxidation of the original ore minerals. This process involves interaction between small parts of each of the three main components of the earth's surface, its crust, its atmosphere and its hydrosphere.

The small scale interactions between these three components have been the focus of scientific study for many years, and are analogous to the data collection period in the evolution of chemistry, but over the last decade many important questions have been posed which demand a vast change in perspective. The expansion of the human population and industrial production has shown that chemical environments on the earth are not infinitely tolerant to change on a human time-scale. Many industrial activities may be slowly but perceptibly changing the operation of the crust-atmosphere-hydrosphere system, for example the huge and continually expanding production of CO_2 from the burning of fossil fuels. How rapidly will the system respond to this influx of CO_2. and what are the probable short and long-term consequences? In attempting to answer these questions environmental chemistry must mature to produce an understanding of the principles controlling earth surface processes which is capable of extrapolation to the future.

The type of problem exemplified by atmospheric CO_2 pollution can only be approached through a change in perspective. The general, global processes at work on the earth's surface must be distinguished from the specific events which occur at isolated points in time and space. An analogy might be made in a visit to the shop-floor of a factory where one would be struck by a sense of confusion and urgency: machinery operates, men move purposefully about their work, materials are transported on fork-lift trucks. Later a visit to an office overlooking the factory site allows one to watch the delivery of basic raw materials and the removal of finished products. Aided by an explanatory block diagram, the factory can be viewed as a sequence of processes which convert raw materials into finished products. Scientists of many different disciplines are now trying to draw together a century of small-scale process observations to derive a block diagram of the earth surface system, or what may be visualised as the earth-air-water factory. Quantitative modelling of the earth-air-water factory can provide the answers to those questions which are concerned with long-term changes induced by human activity.

The Crust

The most common material found beneath the ground is rock. Vegetation, soil, sand and gravel form a thin veneer beneath which bedrock invariably lies. Surfaces and near-surface rocks are similar in composition to those exposed at depths of up to 40 km beneath the continents but further down there is a pronounced compositional break which marks the change from crust to mantle (Fig. 1.1). The compositional break occurs at shallower depths (approximately 6 km) beneath the oceans for reasons which are associated with the formation of the continents and their ocean basins. The thin irregular shell of rock known as the crust constitutes less than 0.0001 % of the earth's volume and its importance lies less in its size than in its accessibility. As a first approximation it is useful to consider the crust as a closed system, i.e. as isolated from the mantle beneath. This is not strictly accurate since volcanic activity may inject mantle material to the surface but such additions are trivial, compared to the total mass of the crust, and are of only localised importance.

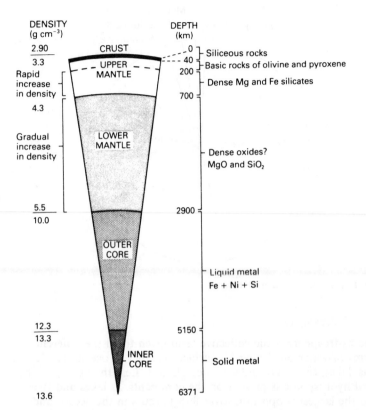

Fig. 1.1 The composition of the earth. Crustal rocks are silica-rich compared to the basic Mg and Fe-bearing rocks in the mantle.

The average composition of the crust, as determined by numerous chemical analyses of rocks, is given in Fig. 1.2. If the compositions of the oceans and atmosphere are included the percentages change only slightly. It is a striking fact that only a few elements are abundant, and most are exceeedingly scarce. Oxygen makes up nearly half the mass of the crust; some of it occurs in the atmosphere, some is combined in water, but the largest fraction is combined with silicon in different proportions to form silicate minerals. Silicon, iron and aluminium together constitute nearly 40% of the crust and another 10% or so is due to calcium, sodium, potassium and magnesium. All the familiar metals, e.g. Cu, Pb, Zn, with widespread domestic and industrial uses are too scarce to be shown individually and merely constitute part of the remaining 1.4%.

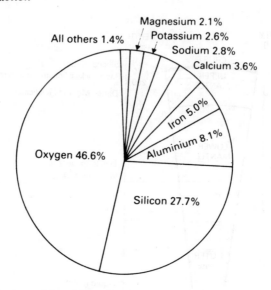

Fig. 1.2 The composition of the crust.

The Hydrosphere

The hydrosphere is the collective term given to all the different forms of water on the earth's surface, whether in oceans, shallow seas, lakes, rivers, groundwaters or glaciers. Less than 0.5 % of the total hydrosphere is present on the continents, as lakes and rivers, and the largest proportion (over 80 %) occurs in the oceans and shallow seas, with virtually all the remainder trapped in buried sediments (Fig. 1.3). About 70 % of the earth's surface is covered by the oceans and thus in terms of both volume and area, the oceans are a major factor in shaping the physical and chemical nature of the earth's surface. For example, climate is modified through the ability of the oceans to absorb solar radiation energy and transport it around the world, as well as through the evaporation/precipitation cycle which is initiated at the air-sea interface. The oceans also play an important role in regulating the abundances of the gases O_2 and CO_2, which are necessary for life processes. Finally, the oceans are linked with the rest of the hydrosphere through the hydrological cycle, in which water evaporated from the oceans into the atmosphere falls as rain or snow on the continents, and returns via rivers and lakes to the oceans again.

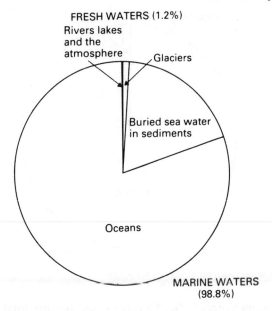

FRESH WATERS (1.2%)
Rivers lakes
and the
atmosphere

Glaciers

Buried sea water
in sediments

Oceans

MARINE WATERS
(98.8%)

Fig. 1.3 The composition of the hydrosphere.

The Atmosphere

The atmospheric component of the earth-air-water factory consists mainly of nitrogen (78 %), oxygen (21 %), argon (0.9 %) and carbon dioxide (0.03 %) mixed with a highly variable proportion of water vapour (Fig. 1.4). The proportions of the major gases in dry air remain fairly constant to altitudes of 100 km, but above this height simple gaseous molecules can break down to ions and free radicals by photodissociation reactions. Oxygen and nitrogen are important biologically and have characteristic cycles of interaction with living organisms. Oxygen is required for the oxidation of food, which provides energy, and is abstracted directly from the atmosphere. In contrast, nitrogen gas is chemically unreactive and the nitrogen required by living organisms to form the amino acids essential for life must be obtained indirectly from the atmosphere by utilising natural processes which convert nitrogen gas into soluble salts. Not only the major gases interact with life processes, for example carbon dioxide is utilised in photosynthesis to make new organic matter with the aid of solar radiation and a catalyst, chlorophyll. The final minor constituent of the atmosphere is the inert gas argon, which owes its relative abundance to production from the breakdown of

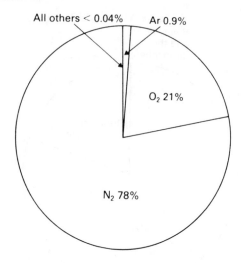

Fig. 1.4   The composition of dry air at sea-level.

the potassium isotope[†] ^{40}K. All other gases together total less than 0.04 % by volume.

The full earth-air-water factory is shown in Fig. 1.5 as a gigantic chemical engineering factory. Some of the more important processes which take place within the factory, i.e. atmospheric chemistry, weathering and soil formation, marine chemistry and the formation of economic deposits in sedimentary rocks, will be discussed.

The Earth-Air-Water Factory

In the earth-air-water factory, as in any man-made factory, energy must be utilised to fuel the successive chemical operations which occur. Many different forms of energy are used in the various operations but all can be considered under one of the following categories: solar radiation; mechanical (kinetic and potential) energy; chemical energy; and the earth's heat content. Solar radiation is clearly the only source external to the earth and, therefore, plays a critical role. The second law of thermodynamics[†] predicts that the earth as a closed system, without external energy inputs, would become rundown and increasingly disordered over a long period of time. That this does not occur is due entirely to the input of solar radiation, consisting chiefly of visible light together with some ultraviolet and infra-red radiation. About 34 % of this is directly

[†] See glossary

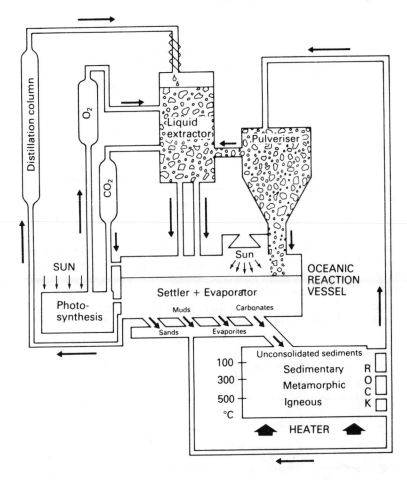

Fig. 1.5 The earth-air-water factory (based on Garrels and Mackenzie, 1971 and Siever, 1974).

reflected back into space by clouds and a further 19 % is absorbed in the atmosphere by ozone, water vapour and carbon dioxide. Only 47 % of the incoming radiation reaches the earth's surface and almost all of this (40 %) is utilised to evaporate water from the hydrosphere, whilst much of the remainder is absorbed by the crust. Only 0.1 % is used in photosynthesis.

The energy used to power the evaporation of water from the hydrosphere reappears as kinetic and potential energy when rain and snow become rivers and glaciers. The small proportion of energy

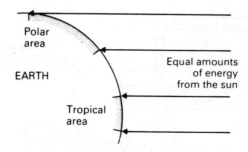

Fig. 1.6 A schematic representation of the relative amounts of solar radiation reaching different latitudes. The same amount of energy reaches all latitudes but in polar regions it is spread over a much larger area.

absorbed by the crust is unevenly distributed over the earth's surface with equatorial regions receiving more energy than the poles, due to more effective heating by vertical, as opposed to oblique, radiation (Fig. 1.6). Since the earth's average temperature is roughly constant, there must be global balance between incoming and outgoing radiation. The rates at which radiant energy enters and leaves the earth are shown in Fig. 1.7 as a function of latitude. More energy

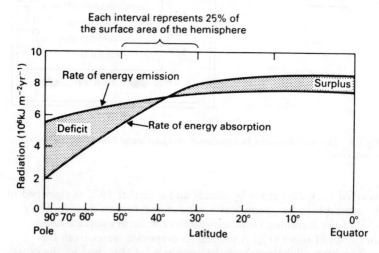

Fig. 1.7 The annual balance between incoming solar radiation and outgoing radiation from the earth. More energy is gained than lost in the tropical regions, and more energy is lost than gained in the polar regions. The latitude scale is spaced so that equal horizontal distances on the graph correspond to equal areas of the earth's surface.

arrives at the equatorial regions than is lost there, but the opposite is true at the poles. Winds and currents are generated to eradicate this imbalance by transferring energy from regions of surplus to regions of deficit. Thus, energy initially absorbed by the crust, e.g. that used to evaporate water, reappears as kinetic and potential energy in the main agents of geological erosion, the winds, tides, glaciers and rivers.

The schematic earth-air-water factory (Fig. 1.5) shows the solar energy input powering the evaporation process operating on the hydrosphere and fuelling photosynthesis. Although photosynthesis uses only a tiny fraction of solar energy it is of fundamental importance to life. Living matter is divided into two categories: producers, e.g. green plants, and consumers, e.g. animals. Green plants utilise chlorophyll and solar radiation to assimilate carbon dioxide for the production of new cell material with the evolution of oxygen, as in the forward reaction

$$6CO_2 + 6H_2O \xrightarrow[\substack{\text{Solar energy} \\ + \text{nutrients} \\ (P, N, Si)}]{\text{Chlorophyll}} C_6H_{12}O_6 + 6O_2. \qquad (1.1)$$

Simple thermodynamics shows that 2879 kJ of energy must be supplied to synthesise 1 mole of carbohydrate, $C_6H_{12}O_6$, that is for this reaction to proceed from left to right under standard conditions (25°C, 1 atm pressure). Solar energy is thus converted into the chemical energy of organic compounds. The oxygen produced is stored in the atmosphere and becomes available to the dependent life systems which utilise the reverse reaction and which constitute the second type of living matter. The animal world, including man, bears an essentially dependent relationship to the plant world, not only in needing oxygen for respiration but also ultimately in needing plant carbohydrates to fuel the reverse reaction. The chemical energy liberated appears as heat and work. The oxygen reservoir is thus maintained by photosynthesis and is utilised in respiratory life systems and oxidative weathering processes. Without living matter, the oxygen content of the atmosphere would be gradually depleted in oxidising the reduced metals exposed on the earth's surface by igneous activity and mountain-building.

The other major gaseous reservoir utilised in the operation of the earth-air-water factory is that storing the carbon dioxide required mainly for photosynthesis and, to a lesser extent, for weathering reactions. The content of this reservoir is partly maintained by the supply of CO_2 from respiration and by volcanic exhalations, but the possible existence of other sources is still being debated (see p. 122). The atmospheric part of the factory lacks a nitrogen reservoir since, although this element, together with phosphorus and sulphur, plays a major role in biological processes, it has only a minor role in the interactions which occur between the crust, atmosphere and hydrosphere. The earth-air-water factory is therefore only a first-order approximation to reality. The gaseous reservoirs interact with the hydrosphere, in particular with the water vapour produced by evaporation of the oceans, by solar radiation. The condensed water vapour dissolves CO_2 to produce acidic solutions which are the weathering agents at work within the liquid extractor plant (see Fig. 1.5). Here, mechanical energy in the form of winds, tides and running water erodes solid rock to produce fine-grained material, which greatly increases the surface area exposed to attack by acids. Soluble components are leached from the rock particles and enter the surface drainage system which eventually reaches the oceans. The solid residues may remain a considerable time at the weathering site, where they constitute a soil system, but ultimately they too are transported into the oceans.

The oceans constitute a huge reaction vessel in which the solid debris and dissolved substances produced by weathering are reacted, evaporated, settled and chemically separated into different components. Thick wedges of sedimentary rocks are formed which slowly sink under their own weight to depths of several kilometres or more, where they are subjected to pressure and heat, with consequent changes in chemical composition. Some sediments are little altered, others are extensively recrystallised to form metamorphic rocks and some may be melted to form magma. The heat for these processes is derived from the radioactive disintegration of uranium, thorium and potassium in the earth's interior. The acid gases and water formed as by-products are exhaled during volcanic eruptions, and so re-enter the atmosphere from which they were removed by weathering. Ultimately the deformed and contorted sedimentary, igneous and metamorphic rocks will also reappear at the earth's surface through tectonic activity and so become exposed once more to weathering reactions.

From this brief description of the earth-air-water factory its cyclic nature is apparent. Two separate but linked cycles can be recognised: that which involves gas-liquid interactions between the atmosphere and hydrosphere and that which involves solid-liquid interactions between the crust and the hydrosphere. The following four chapters will present a more detailed analysis of the processes occurring in these two linked cycles.

2 The Atmosphere

The atmosphere is not only the smallest (mass $= 5.2 \times 10^{18}$ kg) of the three main components in the earth-air-water factory, but it is also the simplest chemically. The gaseous components of the atmosphere play two major roles. They are utilised in biological systems and they determine the initial composition of those solutions which participate in the weathering of the solid components in the crust.

The atmosphere consists mainly of four gases: nitrogen, oxygen, argon and carbon dioxide. Argon, as an inert gas, plays no chemical role in the earth-air-water factory, but the other gases each participate to a greater or lesser extent in the biological and weathering roles. The problem of determining the relative importance of the different roles is one commonly faced by an environmental chemist. Such problems assume particular importance where it is necessary to assess the impact of man's activities on natural processes. A useful approach to dynamic systems in the earth-air-water factory, where the continuous transfer of materials is occurring, is to divide the factory into a series of storage areas, or reservoirs, and to study the flow of materials between these reservoirs. Because the reservoirs are usually represented diagrammatically by boxes, models of this type are sometimes termed 'box models'. Fig. 2.1 shows a simple box model for oxygen, which will be used to demonstrate the properties of box models and the assumptions implicit in their use.

The Oxygen Cycle

Construction of a Box Model

The choice of reservoirs in a box model is determined by the scale of the problem under consideration. The approach adopted here is to study the global behaviour of man-made and natural compounds of environmental significance, and for most purposes atmospheric, crustal and oceanic reservoirs are sufficient. Where one species is converted into another, it may be necessary to subdivide reservoirs and, as in Fig. 2.1, to consider separately the oxygen and carbon dioxide reservoirs in the atmosphere.

12   *The Atmosphere*

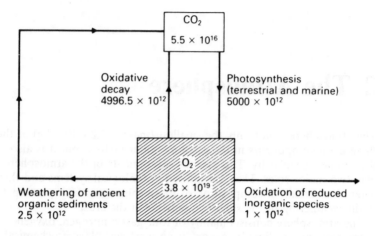

Fig. 2.1 The oxygen cycle in the atmosphere. Reservoir units in moles, fluxes in mol yr^{-1} (after Garrels *et al.*, 1976).

The lower box in Fig. 2.1 contains an estimate of the total oxygen content of the atmosphere, deduced from measurements of its volume and oxygen content. As oxygen is a strong oxidising agent it is inherently unstable with many common earth materials. These oxidation reactions have been assembled into three groups.

(1) The oxidation of organic matter, e.g. coal and oil, contained in ancient sedimentary rocks
(2) The oxidation of reduced inorganic species supplied by geological activity, e.g. Fe^{2+}-bearing minerals exposed by erosion and reduced gases from volcanic activity
(3) The oxidation of modern organic material, e.g. the decay of vegetation in soils, and respiration. These may be termed oxidative decay.

All three reactions consume atmospheric oxygen and involve microorganisms to some extent. Whereas Fig. 2.1 provides a description of the major fluxes to and from the oxygen reservoir, the carbon dioxide reservoir is considered only to the extent of its links with the oxygen reservoir. Thus, processes such as the removal of CO_2 from the atmosphere by rock weathering (page 60) are not shown. Two of the three groups of oxidation reactions generate carbon dioxide or dissolved carbonate species, and their fluxes link the two reservoirs, but the oxidation of reduced inorganic species does not produce CO_2 and merely provides a leak from the oxygen reservoir. Note that the net flux of O_2 out of the atmospheric oxygen reservoir is balanced by photosynthesis, which supplies O_2 from the breakdown of CO_2.

The size of a reservoir can be measured fairly easily but the amounts transferred along particular reaction pathways, or fluxes, are sometimes less easily quantified. Direct measurement of some fluxes is possible, however. For example, rates of photosynthesis can be deduced from laboratory or field studies of organisms using an isotopically labelled, radioactive form of carbon dioxide ($^{14}CO_2$) and measuring its rate of incorporation into new cell material. Knowing the weight of organisms participating in the study, the measured flux can be scaled up to correspond to the rate of photosynthesis by the estimated global population of photosynthetic organisms. Similar field and laboratory studies can be used to estimate rates of oxidative decay, but more indirect methods are necessary for the remaining two fluxes. The oxidation of reduced species includes a variety of contributions but the general approach can be seen by considering two sources: the reduced metals in silicate minerals and the reduced gases exhaled by volcanos. In the former, one method is to estimate annual global rates of erosion and to assume that material eroded has an average crustal abundance of reduced metals. Oxygen consumption is given by balanced chemical equations for the oxidation of reduced metals to common earth surface minerals. For example, the reduced Fe^{2+} in silicate minerals can be oxidised to Fe^{3+} which usually occurs as the hydrated red-brown iron oxide whose composition can be represented by $Fe(OH)_3$.

$$2Fe^{2+}\,(aq) + 5H_2O\,(l) + \tfrac{1}{2}O_2\,(g) \rightarrow 2Fe(OH)_3\,(s) + 4H^+\,(aq)$$

Oxidation-reduction, or redox[*], reactions of this type are explained in the Glossary. However, it can be seen that four Fe^{2+} ions consume one oxygen molecule during oxidation to Fe^{3+} and precipitation as solid $Fe(OH)_3$. The suffixes indicate the physical state of the reactant and product species, i.e. solid (s), liquid (l), gas (g) or dissolved (aq).

Estimates of the reduced species exhaled by volcanoes can be made directly, and, although it is difficult to collect samples of volcanic gases uncontaminated by the atmosphere, reasonably accurate figures are available for the total volume and relative abundance of gases emitted annually by volcanoes. The major reduced gases are H_2S, CO, CH_4 and H_2 and simple chemical equations show their oxygen consumption.

$$2CO + O_2 \rightarrow 2CO_2$$
$$2H_2S + 3O_2 \rightarrow 2H_2SO_3$$

The final major flux of oxygen from the atmosphere represents that consumed in the oxidation of ancient organic matter. Several pathways can provide an estimate of this flux, but only one is discussed here. Geologists have produced fairly reliable estimates of both the annual rate of erosion and the abundance of different rock types exposed on the earth's surface. Assuming that all rocks are eroded at equal rates an estimate may be derived for the volume of oil, coal and other organic-rich sediments destroyed annually by weathering. Chemical analyses can then provide the basis for producing a weighted average organic carbon content of the sediments eroded, and the oxygen consumption estimated by assuming one mole of O_2 is consumed by each mole of organic carbon, represented schematically by the simplest carbohydrate CH_2O.

$$CH_2O + O_2 \rightarrow CO_2 + H_2O$$

Estimates of fluxes by these indirect methods may be subject to quite large errors, although, generally, the largest fluxes are quite accurately known and, therefore, the large errors in the lesser-known and smaller fluxes have relatively little overall impact on the box model. The advantage of the box model approach is that it clearly shows the susceptibility of a reservoir to change and identifies which process is most likely to cause that change.

The complete box model for O_2, together with flux magnitudes, is shown in Fig. 2.1. Its cyclic arrangement should be noted. For example, atmospheric O_2 may be transferred around the cycle, being consumed by oxidative decay and then regenerated from CO_2 by photosynthesis. A closed cycle of this type is assumed to be in a steady state, i.e. the individual reservoirs do not alter in size, which implies that the total rates of addition to or removal from a reservoir are exactly equal and do not alter. Although this is not always true, it appears to be a reasonable assumption for many natural materials in the earth-air-water factory. The arguments on which this assumption is based rest largely on the fact that if cycling did not closely approximate to a steady state there would be observable changes in the compositions of the crust, oceans and atmosphere. Scientific results generally show that such changes have not occurred during the twentieth century, although one important exception is the increased CO_2 content of the atmosphere. Further, the chemical similarities between modern and ancient sediments, and the known fossil record, imply that no significant changes in the chemical composition of the crust, atmosphere or oceans have occurred. A man transported back 300 million years in time would find the basic necessities of life unaltered. There would be water to drink, food to eat and air to breathe and, although there would be great changes in the structure and diversity of life-forms, there would be no major changes in the chemical operation of the earth-air-water factory.

Residence Time and the Assessment of Anthropogenic Influences

In order to assess the susceptibility of a reservoir to change the residence time for any component (*c*) in the reservoir must be defined:

$$\text{Residence time} = \frac{\text{Total content of } c \text{ in reservoir}}{\text{Total rate of input or output of } c}$$

The significance of residence time can be seen in the following analogy. A bath contains 500 l of water. If the plug is removed and the tap turned on sufficiently to keep the volume constant, the reservoir of water in the bath will be in a steady state. If the rate of inflow is 50 l min^{-1}, the residence time, which is the time taken to add or subtract an amount of water equal to that originally in the reservoir, is 10 minutes. Computation of residence times enables two types of environmental problems to be identified. Where the residence time of a pollutant is large it may remain in the environment for a considerable time before becoming inactivated, and where residence times are small a reservoir is relatively sensitive to changes in the rates of inflow and outflow as a result of man's activities.

Residence time also plays an important role in determining the spatial variability of atmospheric gases on a global scale. By using the standard deviation (σ) of the average mixing ratio (weight of gas : weight of air) as a measure of spatial variability, Junge (1974) was able to show a linear, but inverse, relationship between log σ and residence time (Fig. 2.2). This relationship allows the significance of transport mechanisms in the atmosphere to be evaluated. A careful study of Fig. 2.2 shows that where log σ is smaller than approximately 0.3 the residence time is longer than half a year. In this case, which includes all the major components of the atmosphere, mixing occurs between the southern and northern hemispheres sufficiently rapidly for the gas to be at an essentially uniform concentration, on a global scale. Details of atmospheric transport mechanisms are not then important in determining the way such constituents are dispersed. The reverse applies where log σ is greater than 0.3 and average concentration is dependent on geographical location, climate and distance from the source area.

In the oxygen cycle the residence time of atmospheric oxygen is 7600 years and it is clear that even large changes in the rates of outflow do not deplete the reservoir to any significant extent on a human time-scale. This is shown clearly in Broecker's study of man's

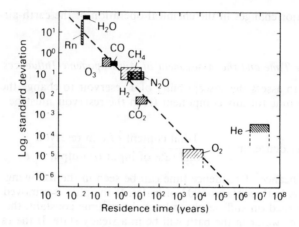

Fig. 2.2 The relationship between spatial variability of atmospheric gases, as measured by log standard deviation of mixing ratio, and residence time. Atmospheric mixing occurs sufficiently rapidly to give uniform global concentrations of those gases whose residence times are greater than 0.5 yr (from Junge, 1974).

influences on atmospheric oxygen levels (1970). The major potential human interference to the oxygen cycle is through the burning of fossil fuels, which consumes oxygen. In 1978 the world-wide consumption of coal and oil was estimated to involve the oxidation (burning) of 400×10^{12} moles of carbon. During combustion each mole of carbon consumes one mole of oxygen, but this amount of oxygen constitutes only 0.001 % of that in the atmospheric reservoir. Man has recovered altogether about 10^{16} moles of fossil carbon from the crust, which probably represents about 4 % of our total fossil fuel reserves (25×10^{16} moles of organic carbon). Burning all known fossil fuel reserves would, therefore, consume less than 1 % of the oxygen in the atmosphere. Unfortunately the atmospheric reservoir is sensitive to the effects of fossil fuel combustion in other ways, e.g. the addition of CO_2 (see page 33), but at least the depletion of O_2 by this process is unlikely on any reasonable time-scale.

One other aspect of the oxygen cycle is worthy of comment. Its most important links are to the biological cycles through respiration and photosynthesis and this has lead to speculation that if photosynthesis ceased the atmosphere would be rapidly depleted of oxygen and human life would cease. Jacques Cousteau, in a letter from the Cousteau Society, has predicted the events following abrupt termination of marine photosynthesis, perhaps as a result of catastrophic pollution. 'With life departed, the ocean would become in effect, one enormous cesspool. Billions of decaying bodies, large and small, would create such a stench that man would be forced to leave the coastal regions . . . . Then would be visited the final plague, anoxia . . . . And so man would finally die, slowly gasping out his life on some barren hill. He would have survived the oceans for perhaps 30 years.' Cousteau's description may be accurate but clearly his estimate of the time-scale of man's survival is much too short. The 7600 year residence time of oxygen in the atmosphere represents the minimum time before the O_2 reservoir can become exhausted, assuming no regeneration by either terrestrial or marine photosynthesis. In fact the terrestrial photosynthesis of oxygen would probably continue in these circumstances and could supply approximately 50% of the total photosynthetic flux. Thus the time taken to deplete the atmosphere of O_2 would be represented by the residence time of O_2 with respect to its net flux used in global oxidative decay and generated by terrestrial photosynthesis. This would give a time of 15 000 years, which even then represents the minimum period for oxygen removal from the atmosphere since rates of oxidative decay would probably decline rapidly from their present values without new organic material being generated by marine photosynthesis. A substantially longer period than 15 000 years would, therefore, be required before complete depletion occurred. In either case the real problem would be starvation, rather than anoxia, since man is ultimately dependent on photosynthetic plants as a food source.

The Nitrogen Cycle

The oxygen cycle is simple, but that for nitrogen, the other major atmospheric gas, is remarkably complicated (Fig. 2.3). This in part arises from introducing more detail by considering the form in which nitrogen is present in each reservoir. Thus different inorganic and organic nitrogen species can be identified in each of the gas, liquid and solid phases which occur in the earth-air-water factory. Up to a point refinements of this sort improve our understanding of nitrogen behaviour but they do require more detailed analytical information which is often not available.

   The variety of species in the nitrogen cycle is large, due to the wider range of oxidation numbers [+] of nitrogen (commonly -3 in NH_3, 0 in N_2, $+1$ in N_2O, $+3$ in NO_2^- and $+5$ in NO_3^-) compared to oxygen (0 in O_2 and -2 in H_2O). Some grouping is necessary to simplify the diagram and, in the atmosperic reservoir, species with oxidation numbers greater than $+1$ or less than -1 (NH_3, NH_4^+, NO_2) are separated from species with oxidation numbers -1, 0, $+1$ (N_2, N_2O). In the crustal and oceanic reservoirs the distinction is drawn on the nature of the nitrogen-bearing phase, i.e. whether it is organic or inorganic. .

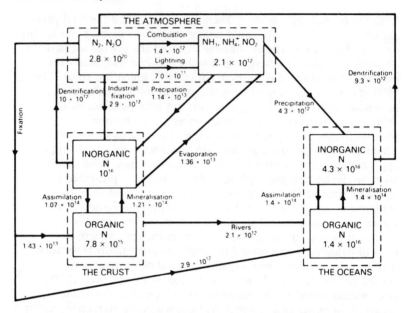

Fig. 2.3   The nitrogen cycle. Reservoir units in moles of nitrogen (expressed as N), fluxes in mol yr^{-1} (after Liu *et al.* 1977).

Another important feature of the nitrogen cycle is that a relatively small number of processes, i.e. mineralisation, fixation, assimilation and denitrification (Table 2.1), are responsible for most of the fluxes between the different species reservoirs. Mineralisation represents the conversion of nitrogen in organic compounds, chiefly amino acids, into inorganic forms of nitrogen e.g. NH_3, NO_3^-, NO_2^-. The process operates when microorganisms decompose the remains of dead plants and animals in both terrestrial, e.g. soil, and marine environments. The most critical process in the terrestrial part of the cycle is fixation, by which gaseous N_2 from the atmosphere is converted into a form which can be utilised by plants and animals. Although nitrogen constitutes 79 % of the atmosphere it cannot be used directly by the large majority of living things, instead it must first be fixed by specialised organisms, such as blue-green algae. After their death the organisms which can fix nitrogen are themselves mineralised to inorganic nitrogen species, and this is the major pathway by which the inorganic nitrogen species necessary to support other life systems are synthesised. This conversion of inorganic nitrogen to organic nitrogen is the process of assimilation. The small subcycle between assimilation, death and mineralisation

Table 2.1 Chemical fluxes in the nitrogen cycle

Mineralisation

Conversion of organic N to inorganic N

$CH_2NH_2COOH + 3/2O_2 \rightarrow 2CO_2 + H_2O + NH_3$
  Glycine
  (amino acid)
NH_3 can then be oxidised to NO_3^- and NO_2^-

Fixation

Conversion of atmospheric N_2 to organic N

$N_2 + 3H_2 \rightarrow 2NH_3$
$2CO_2 + H_2O + NH_3 \rightarrow CH_2NH_2COOH$

Assimilation

Conversion of inorganic N to organic N

$4NO_3^- + 8H_2O \rightarrow 4NH_3 + 8O_2 + 4OH^-$
$NH_3 + 2CO_2 + H_2O \rightarrow CH_2NH_2COOH$

Denitrification

Conversion of inorganic N to atmospheric N_2

$4NO_3^- + 2H_2O \rightarrow 2N_2 + 5O_2 + 4OH^-$

represents the recycling of nitrogen within the soil system in which most animal and plant life, including all agriculture, participates. It is critically dependent on fixation as the major independent natural source of available inorganic nitrogen. The industrial fixation flux represents man's attempts to supplement the available inorganic nitrogen in the soil system by synthesising ammonia from N_2, H_2 and CH_4 in the Haber process and thus forming salts for fertilisers. Before the development of the Haber process the major source of fertiliser was natural deposits of nitrates. Natural nitrate deposits would soon have become exhausted and modern agricultural methods, critically dependent on fertilisers to maintain available inorganic nitrogen levels, could not have been developed without industrial processes to fix gaseous nitrogen.

There is one other important flux of inorganic nitrogen into the terrestrial environment which arises from the scavenging of soluble gases, mainly NH_3, NH_4^+ and NO_3^-, by precipitation from the atmosphere by rain. These species are created naturally by lightning and biological activity, and artificially from the combustion of fossil fuels.

The description of important fluxes in the nitrogen cycle has so far concentrated on the terrestrial part of the cycle, but the same processes of fixation, precipitation, denitrification, assimiliation and mineralisation are also present in marine environments. The same reservoirs can also be recognised.

Most attention, however, is usually focused on the industrial fixation of nitrogen in agriculture, because this constitutes a comparatively large interference in the natural cycle of nitrogen. Since 1950 the amount of nitrogen annually fixed as fertilisers or as nitrogen compounds (e.g. NH_4NO_3) has increased approximately fivefold, until it now equals the amount that was fixed by all terrestrial ecosystems before the advent of modern agriculture. Addition of fertiliser nitrogen to the soluble inorganic nitrogen reservoir, aided by the precipitation of fossil fuel combustion products, has had the consequence of removing the steady state condition from the inorganic nitrogen reservoir. Removal processes such as denitrification do not appear to be keeping pace and the excessive run off of nitrogen into streams and lakes can result in blooms of algae, whose intensified activity deplete the other important nutrients, e.g. P, in the water. Starvation and death of the algae population then follows and the decay of their cell material by aerobic metabolism (see page 140) depletes the water of oxygen, such that fish and other oxygen-dependent organisms are destroyed. This condition is known as eutrophication and will be discussed further (see page 90).

In considering the nitrogen cycle as a whole, one may wonder how it is that some organisms are able to oxidise nitrogen compounds whereas other organisms, often in the same environment, have life systems based on their ability to reduce nitrogen compounds. Apart from photosynthetic organisms which obtain their energy from solar radiation, all living organisms depend for their energy on chemical transformations. These transformations typically involve the oxidation of one compound and the reduction of another, although in some cases the compound being oxidised and that being reduced may be different molecules of the same substance. Nitrogen can be cycled because the reduced inorganic compounds of nitrogen can be oxidised by atmospheric oxygen with a useful yield of energy to microorganisms carrying out this reaction. Similarly, in the absence of oxygen, the oxidised compounds of nitrogen can act as oxidising agents for the consumption of organic compounds, again with a useful yield of energy. Nitrogen is able to play its complex role in life processes because it has a large number of oxidation states[†] the transfers between which may be energetically favourable in certain circumstances.

Thermodynamics and the Prediction of Chemical Reactions

The discussion of chemical reactions in terms of their energy transfers is known as thermodynamics. The study of thermodynamics can provide us with a valuable predictive tool to judge which chemical reactions are feasible, simply by considering the change in energy between the reactants and their products. Reactions in which there is a net liberation of energy can occur spontaneously but reactions in which energy is consumed require an external energy source. This applies both to biological processes and to inorganic reactions. For example, photosynthetic organisms utilise solar radiation to convert carbon dioxide and water to carbohydrate using a chlorophyll catalyst:

$$6CO_2 + 6H_2O \xrightarrow[\text{Solar energy} + \atop \text{nutrients (P, N, Si)}]{\text{Chlorophyll}} C_6H_{12}O_6 + 6O_2 \qquad \begin{array}{c} 2879 \text{ kJ} \\ \text{required} \end{array}$$

However, the reverse reaction, representing the microbiological decay of dead organic matter occurs spontaneously:

$$6O_2 + C_6H_{12}O_6 \xrightarrow{\text{Microorganisms}} 6CO_2 + 6H_2O \qquad \begin{array}{c} \text{Nutrients and} \\ 2879 \text{ kJ} \\ \text{liberated} \end{array}$$

The energy liberated is written on the right side of the equation and conventionally given a negative sign. Where energy is required to make a reaction occur from left to right, the energy requirement is expressed as a positive quantity. Energy liberated often appears as heat, e.g. the temperature reached as a result of microbiological decay in a garden compost heap may reach 40° C, or work. Some chemical reactions are assisted by the presence of a catalyst which plays a role shown schematically in Fig. 2.4. Catalysis may aid both spontaneous and non-spontaneous reactions. In both cases there may be an energy barrier, the activation energy, which prevents the transformation occurring. The size of the energy barrier is reduced with a catalyst, which forms an activated intermediate with the reactants. The intermediate then breaks down spontaneously to form the products. The two cases differ simply in the net energy change between the reactants and products, which cannot be altered by the presence of a catalyst. Reactions involving catalysts may be written as simple transformations between starting materials and products, but are much more complicated. For example, nitrogen fixation (page 19) is written as

$$N_2 + 3H_2 \rightarrow 2NH_3 \qquad\qquad +53.6 \text{ kJ}$$

but is in fact a multi-stage process catalysed by enzymes. In these circumstances the source of external energy must ultimately be solar

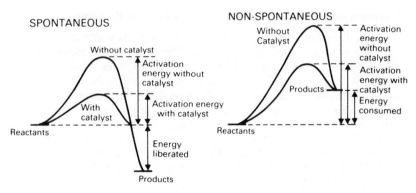

Fig. 2.4 The role of a catalyst. The energy barrier which hinders reaction progress (the activation energy) is reduced by the presence of a catalyst. The catalyst has no effect on the net energy change in the overall reaction.

radiation but the reaction mechanisms are poorly understood.

In the nitrogen cycle (Fig. 2.3) there are two non-biogenic oxidation-reduction processes in which atmospheric N_2 may participate: combustion and lightning may each oxidise N_2 to the higher gaseous oxides. The latter reaction provides a good illustration of the predictive powers of thermodynamics in quantifying the amounts of energy used or liberated in a reaction. Since oxygen is a powerful oxidising agent why is it only possible to oxidise nitrogen to nitric oxide and nitrogen dioxide with the aid of an energy input from lightning?

$$N_2 + O_2 \rightarrow 2NO$$
$$2NO + O_2 \rightarrow 2NO_2$$

The general approach is outlined in the Glossary, with necessary definitions of thermodynamic terms and units. Tables of thermodynamic data show that the free energy[†] of formation, $\Delta G^{\ominus}$, of NO is $+86.6$ kJ mol^{-1}, so the free energy change on oxidation to nitrogen dioxide is calculated as follows:

$$N_2 + O_2 \rightarrow 2NO$$
$$\Delta G^{\ominus} = \Delta G^{\ominus} \text{ (products)} - \Delta G^{\ominus} \text{ (reactants)}$$
$$= 2 \times 86.6 - 0 - 0 = +173.2 \text{ kJ}.$$

The positive value indicates that the reaction will not proceed from left to right unless there is an external input of energy. In the upper regions of the atmosphere, high temperatures and electric discharges are capable of supplying the energy needed, yet the production of nitric oxide is minimal. The reason can be seen from the Law of Mass Action[†] for this oxidation reaction:

$$K_E = \frac{[NO]^2}{[O_2][N_2]}$$

The value of K_E can then be derived from the free energy change:

$$\log_{10} K_E = \frac{-\Delta G^{\ominus}}{5.707}$$

from which it can be calculated that $\log K_E = -30.3$ and $K_E = 10^{-30.3}$.

This value of the equilibrium constant is so low that the amounts of nitric oxide formed must be very small in comparison with the concentrations of nitrogen and oxygen. The actual amounts of nitric oxide present in an equilibrium situation can be calculated by inserting the values for the partial pressures† of N_2 and O_2 into the equilibrium expression.

$$[NO]^2 = K_E[N_2][O_2]$$

As nitrogen constitutes about four-fifths of the atmosphere and oxygen one-fifth, the partial pressures of N_2 and O_2 are 0.8 and 0.2 respectively. Therefore,

$$[NO]^2 = 10^{-30.3} \times 0.8 \times 0.2$$
$$[NO] = 2.8 \times 10^{-16}.$$

Therefore the equilibrium atmospheric concentration of nitric oxide is minimal. Furthermore the forward reaction is so slow that equilibrium is never approached except perhaps in the upper atmosphere where the formation of nitric oxide occurs more rapidly through a photochemical mechanism. Once formed the nitric oxide oxidises spontaneously to nitrogen dioxide:

$$2NO + O_2 = 2NO_2 \quad \Delta G^{\ominus} = -69.5 \text{ kJ}.$$

The resulting nitrogen dioxide can be leached from the atmosphere by water vapour and the overall process constitutes one of the few ways, apart from the fixation-mineralisation pathway, in which inorganic nitrogen salts can be synthesised naturally from atmospheric nitrogen. The nitrogen cycle in Fig. 2.3 shows the precipitation flux of nitrogen salts from the atmosphere to the earth reservoir of soluble inorganic nitrogen. It is interesting that the same overall process also occurs in high temperature combustion and the resulting nitrogen dioxides in city air may give rise to the photochemical smogs of cities such as Los Angeles (page 43).

Chemical Interactions between the Atmosphere and the Hydrosphere

Within the earth-air-water factory, interactions between the atmosphere and hydrosphere control the composition of the aqueous solvent utilised in the liquid extractor plant to weather rock minerals. The example of interactions between different atmospheric components discussed in the previous section showed how simple chemical principles can be used to predict the nature and magnitude of earth processes. The same principles can also be applied to reactions between the major gas reservoirs and the hydrosphere. The major gases are all soluble in water to an extent which is measured by their Henry's law constants, K_H,[†] (Table 2.2), which are equilibrium constants for reactions of the following type:

$$O_2(g) \rightleftharpoons O_2(aq)$$

$$K_H = \frac{[O_2(aq)]}{[O_2(g)]} \tag{2.1}$$

The amount of gas in solution is a function of the partial pressure of that gas, in atmospheres, which is proportional to the mole fraction [†] of that gas. Oxygen, for example, represents 20% of the air and exerts a pressure of 0.2 atm at the earth's surface. By deriving partial pressure values and using the values of K_H in Table 2.2, the concentrations of the major gases dissolved in pure water can be calculated. It should be noted that values for dissolved concentrations cluster much closer together than do the atmospheric concentrations, because the less common gases are more soluble. This might be why they are less common.

Table 2.2   Solubilities of the major gases in pure water at 20°C

Gas	K_H mol 1^{-1} atm^{-1}	Dissolved concentration mol 1^{-1}
Nitrogen	6.79×10^{-4}	5.3×10^{-4}
Oxygen	1.34×10^{-3}	2.8×10^{-4}
Argon	1.43×10^{-3}	1.3×10^{-5}
Carbon dioxide	3.79×10^{-2}	1.2×10^{-5}

Carbon dioxide is more complicated than the other gases as it reacts with water in the following series of reactions. Dissolved carbon dioxide is written as H_2CO_3, although in fact the species $CO_2(aq)$ is more abundant than H_2CO_3. Chemically and thermodynamically, however, there is no need to distinguish between the two.

$$H_2CO_3(aq) \rightleftharpoons HCO_3^-(aq) + H^+(aq) \tag{2.2}$$

$$HCO_3^-(aq \rightleftharpoons CO_3^{2-}(aq) + H^+(aq) \tag{2.3}$$

The equilibrium constants for these reactions are

$$K_{(2.2)} = \frac{[H^+][HCO_3^-]}{[H_2CO_3]} = 4.5 \times 10^{-7}$$

$$K_{(2.3)} = \frac{[H^+][CO_3^{2-}]}{[HCO_3^-]} = 4.7 \times 10^{-11}$$

Rewriting the second of these relationships gives

$$\frac{[CO_3^{2-}]}{[HCO_3^-]} = \frac{K_{(2.3)}}{[H^+]} = \frac{4.7 \times 10^{-11}}{[H^+]}$$

If $\quad [HCO_3^-] = 10[CO_3^{2-}]$

then $\quad [H^+] \quad = 4.7 \times 10^{-10}\ \text{mol}\,l^{-1}$

and $\quad$ pH $\quad = 9.33$

For all pH values less than 9.3, $[HCO_3^-]$ is at least ten times greater than $[CO_3^{2-}]$ (Fig. 2.5). The pH of most natural systems is less than 9.3 so $[HCO_3^-] \gg [CO_3^{2-}]$ and $[CO_3^{2-}]$ may be ignored to a first approximation. Hence the total dissolved carbon dioxide, T_{CO_2}, can be written

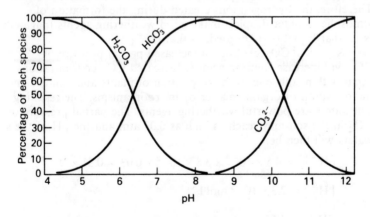

Fig. 2.5 Relationship between dissolved carbonate species and pH. The undissociated acid $[H_2CO_3] > [HCO_3^-]$ for pH < 6.4; $[H_2CO_3] < [HCO_3^-] > [CO_3^{2-}]$ for pH 6.4 – 10.3; $[CO_3^{2-}] > [HCO_3^-]$ for pH > 10.3.

$$T_{CO_2}(aq) = [H_2CO_3(aq)] + [HCO_3^-(aq)].$$

By analogy with (2.1), the Henry's law expression is

$$K_H = \frac{[H_2CO_3]}{[CO_2(g)]} \tag{2.4}$$

and from Equation (2.2)

$$[HCO_3^-] = K_{(2.2)}\frac{[H_2CO_3]}{[H^+]} \tag{2.5}$$

Combining (2.4) and (2.5)

$$[HCO_3^-] = \frac{K_{(2.2)} \times K_H \times [CO_2(g)]}{[H^+]}$$

However, in the solution of carbon dioxide in pure water, all the H^+ ions arise from the dissociation of H_2CO_3 so there must be equal concentrations of H^+ and HCO_3^- ions. The partial pressure of CO_2 in the atmosphere is 0.00032, therefore

$$\begin{aligned}
[H^+]^2 &= K_{(2.2)} \times K_H \times [CO_2(g)] \tag{2.6}\\
&= 4.5 \times 10^{-7} \times 3.8 \times 10^{-2} \times 0.00032\\
&= 5.5 \times 10^{-12}\\
[H^+] &= 2.3 \times 10^{-6} \text{ mol l}^{-1}\\
\text{pH} &= -\log_{10}[H^+] = 5.64
\end{aligned}$$

This is an important result. The pH of water in equilibrium with atmospheric CO_2 is 5.6 approximately, which is an acidic solution. Therefore, the hydrogen ions created during the formation of atmospheric water droplets make good weathering agents (page 60). Note that Equation (2.6) predicts that $[H^+]$ = $\sqrt{K_{(2.2)}K_H[CO_2(g)]}$ and an increased partial pressure of CO_2 will increase solution acidity. Increased CO_2 concentrations occur within soils due to the respiration of plants and animals and the oxidation of organic matter by microorganisms. The resulting soil waters are powerful weathering agents. The partial pressure of CO_2 in the soil may reach as high as 0.03 atm and the pH of the soil waters will then be;

$$[H^+]^2 = 4.5 \times 10^{-7} \times 3.8 \times 10^{-2} \times 0.03 = 0.51 \times 10^{-9}$$

$$[H^+] = 2.3 \times 10^{-5} \text{ mol l}^{-1}$$

$$\text{pH} = 4.65$$

This indicates that soil waters may be as much as one pH unit lower than the same waters in equilibrium with the atmosphere.

The same treatment can now be applied to another gas, sulphur dioxide, which is normally present only in trace amounts in the atmosphere but which may be found in increased concentrations near volcanoes, hot springs or near large cities. Oxides of sulphur result from the burning of fossil fuels, possibly in extreme cases reaching a concentration of 10^{-6} atm. Dissolution of SO_2 in atmospheric water occurs in a series of dissociation steps which are analogous to the CO_2 case:

$$SO_2(g) + H_2O(l) \rightleftharpoons H_2SO_3(aq) \qquad K_H = 1.24 \qquad (2.7)$$

$$H_2SO_3(aq) \rightleftharpoons H^+(aq) + HSO_3^-(aq) \quad K_{(2.8)} = 1.27 \times 10^{-3} (2.8)$$

$$HSO_3^-(aq) \rightleftharpoons H^+(aq) + SO_3^{2-}(aq) \quad K_{(2.9)} = 6.24 \times 10^{-8} (2.9)$$

Therefore,

$$K_H = \frac{[H_2SO_3]}{[SO_2(g)]}$$

$$K_{(2.8)} = \frac{[H^+][HSO_3^-]}{[H_2SO_3]}$$

$$K_{(2.9)} = \frac{[H^+][SO_3^{2-}]}{[HSO_3^-]}$$

An identical series of steps to those used for CO_2 can be used to show that

$$\frac{[SO_3^{2-}]}{[HSO_3^-]} \leqslant 0.1 \quad \text{if} \quad \frac{6.24 \times 10^{-8}}{[H^+]} \leqslant 0.1$$

i.e. $\quad [H^+] \geqslant 6.24 \times 10^{-7} \text{ mol } l^{-1}$

Therefore, HSO_3^- dominates over SO_3^{2-} for all pH $\leqslant 6.2$ and $[H^+]^2 = K_H K_{(2.8)}[SO_2(g)]$.

A typical partial pressure of SO_2 in a moderately polluted urban environment is 10^{-7} atm and the pH of atmospheric water droplets in equilibrium with this SO_2 concentration is, as follows:

$$[H^+]^2 = 1.24 \times 1.27 \times 10^{-3} \times 10^{-7}$$

$$[H^+] = \sqrt{1.57 \times 10^{-10}}$$

$$\doteq 1.25 \times 10^{-5} \text{ mol } l^{-1}$$

$$\text{pH} = 4.9$$

The Henry's law constant and the dissociation constant for the dissolved gas are so much greater for sulphur dioxide than for carbon dioxide that even at comparatively low concentrations (10^{-7} atm compared to $10^{-3.5}$ for CO_2) the resulting solutions of the gas are much more acidic. Thus the dissociation of H_2SO_3, into H^+ and HSO_3^-, occurs more readily than the dissociation of H_2CO_3 and solutions of the former contain more hydrogen ions and are said to be stronger acids.

The above chemistry gives a theoretical basis for understanding the powerful leaching action of rain falling in the vicinity of industrial sources of sulphur dioxide, which is one of the major causes of eroded stonework in large cities. In this way man's activities exert a significant and novel effect on weathering processes.

As well as the acidification role of the atmosphere, there is also an oxidation role, and together these two mechanisms constitute the principle ways in which the atmosphere and hydrosphere interact within the earth-air-water factory. The oxidative role arises because water droplets in the atmosphere contain not only dissolved acidic gases, but also dissolved oxygen. One aspect of this oxidation role is examined here by expanding the SO_2 case. The HSO_3^- ions which result from the solution of SO_2 in rainwater can be oxidised as follows:

$$HSO_3^- (aq) + \tfrac{1}{2}O_2 (g) \rightarrow HSO_4^- (aq)$$

The hydrogen ion in HSO_4^- is only bound very loosely to the SO_4^{2-}, and dissociates as follows:

$$HSO_4^- \rightarrow H^+ (aq) + SO_4^{2-} (aq)$$

The separate, but linked, dissolution and oxidation steps have effectively produced sulphuric acid from sulphur dioxide, oxygen and water:

$$SO_2(g) + H_2O(1) + \tfrac{1}{2}O_2(g) \rightarrow H_2SO_4(aq) \rightarrow 2H^+(aq) + SO_4^{2-}(aq)$$

Strong acids like H_2SO_4 are not only the product of man's intervention in the global factory but can also result where beds of sulphide-rich rocks become exposed to the atmosphere. It is, however, relatively rare for mineral reactions involving sulphur species to be quantitatively important in the earth-air-water factory, although the acidity resulting from the dissolution and oxidation of gaseous sulphur dioxide may have an important, but localised, economic impact. For example, acidic precipitation resulting from SO_2 emissions by industrial Europe may have caused widespread death of fish in parts of Scandinavia (page 60).

Rainwater typically has a pH between 5 and 6, indicating a predominant control by carbonate equilibria. Unusually acidic values arise from SO_2 emissions, and values higher than the theoretical pH of 5.6 may also arise due to the occurrence of alkaline materials in the atmosphere. The most frequently encountered gaseous alkali is ammonia which originates from decaying protein material and which is also readily soluble ($K_H = 58$ mol l^{-1} atm^{-1}) in atmospheric water droplets. Neutralisation of the sulphuric acid contained within the droplet occurs and ammonium sulphate is formed, in this way the soil system in effect receives ammonium sulphate, a useful fertiliser, via atmospheric precipitation. It is thought that the enhanced levels of atmospheric sulphur which result from coal-burning may make an important contribution to soil fertility. Fertilisers containing sulphate are often necessary to counteract sulphur deficiency in non-coal burning areas, but never in industrial areas. This form of pollution is, thus, ambivalent in its environmental impact.

Atmospheric Water and the Water Cycle

The moisture content of the atmosphere is mainly derived from the evaporation of sea water, but a little comes from the evaporation of lakes, rivers, soil moisture and vegetation. The moisture content of the atmosphere at any particular location may vary considerably, since water is continually being added to the air by evaporation and removed by condensation as clouds, rain, snow and fog. A box model of the water cycle in the earth-air-water factory shows how this variability arises.

Fig. 2.6 shows the flux of water between the three main reservoirs in the earth-air-water factory. Over the land area the average annual precipitation is 71 cm yr^{-1} whilst evaporation rates are significantly lower at 47 cm yr^{-1}. The reverse situation exists over the ocean,

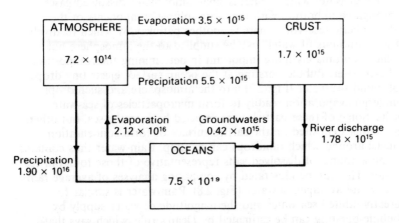

Fig. 2.6 The water cycle. Reservoir units in kg moles, fluxes in kg mol yr^{-1}.

where precipitation is about 110 cm yr^{-1} and 120 cm yr^{-1} are evaporated. These figures reflect the fact that the land gains water over the oceans. The surface area of the oceans is over twice that of the land area, therefore evaporation rates from the ocean are approximately half the precipitation rates on land. Water lost from the oceans by evaporation to the atmosphere is recycled rapidly back to the oceans by atmospheric precipitation and continental run off. Because the atmosphere is such a small reservoir compared to the crust and the hydrosphere, the rate of water transfer through it takes place relatively rapidly. The residence time of water in each of the reservoirs can be derived as follows:

$$\text{Residence time} = \frac{\text{Total } H_2O \text{ content of reservoir}}{\text{Rate of } H_2O \text{ input or output}}$$

The total amount of water in the atmosphere is approximately 7.2 × 10^{14} moles and the rate of input is equal to the net loss by the oceans, which is also the same as the net gain by the continents (2.2 × 10^{15} mol yr^{-1}). Simple division gives the residence time in the atmosphere as 0.03 years, or 11 days. By contrast the residence time in the oceans (volume = 9.5 × 10^{19} moles) is 3500 years and on the continents (water volume 1.7 × 10^{15} moles) is approximately one year. This shows that with small reservoirs such as the atmosphere, even low to moderate rates of transfer allow all the material in that reservoir to be replaced within very short lengths of time. Thus the water content of the atmosphere may show large spatial and temporal variations, as shown in Fig. 2.2.

Evaporation from the hydrosphere provides an essentially pure source of water which contains small amounts of dissolved gases through equilibration with the more soluble components of the atmospheric reservoir. There is another pathway by which relatively trivial amounts of water may be supplied to the atmosphere but which is ultimately more important in determining the composition of rainwater. Bubble-bursting at the ocean surface ejects tiny drops of liquid sea water (Fig. 2.7) into the atmosphere and such drops undergo evaporation readily to form microparticles of sea-water salts. Some of these particles are recycled back to the sea, but others are either deposited on the earth's surface or form condensation nuclei around which raindrops can develop. Rain-water thus contains small amounts of dissolved salts representative of those found in sea water. This can be illustrated by comparing analyses of average rain-water and average sea water (Fig. 2.8). Rain-water is similar to greatly diluted sea water and the magnitude of water supply by bubble-bursting can be estimated by Dean's rule, which says that rain-water is made by mixing 1 l of distilled water with 1.5 ml of sea

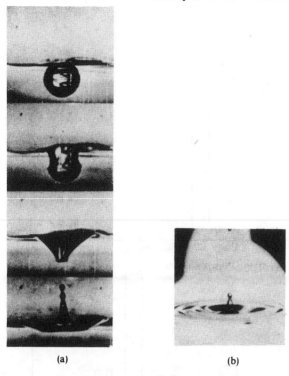

(a)   (b)

Fig. 2.7 Composite view of high-speed motion pictures illustrating some of the
stages in the ejection of drops during bubble bursting: a, the time interval over the
sequence is 2.3 ms and the angle of view is horizontal; b, an oblique view of the final
stages of bubble bursting (courtesy of D. C. Blanchard).

water. Bubble-bursting, therefore, supplies no more than 0.15 % of
the water contained in the atmosphere.

Since Na^+ and Cl^- are the dominant ions in sea water it is not
unexpected that these ions are the most abundant ones contributed
to the hydrosphere by rain-water. The marine component of rain-
water is fairly evenly distributed over the central parts of the
continents but is rather larger towards the coasts. Fig. 2.9 shows the
Cl^- content of rain-water as a function of distance from the coast. It
is apparent that within a few tens of kilometres off the coast the Cl^-
concentration, as well as Na^+ and other ions from sea water, drops
off rapidly to reach a steady background level which remains fairly
constant inland of the coastal strip. This result implies that a
significant proportion of the sea salt injected into the atmosphere by
bubble-bursting at the sea surface is rained out within a short
distance inland from the coast.

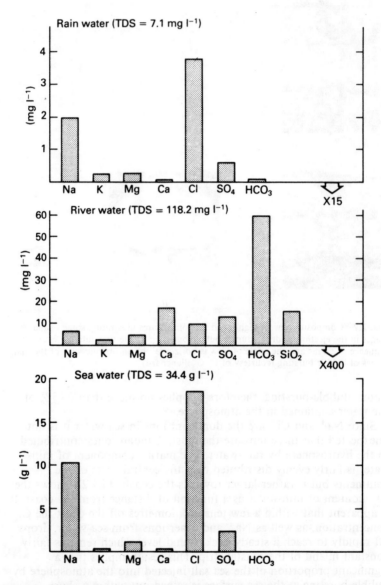

Fig. 2.8 The chemical composition of seawater, river water and rainwater (TDS = total dissolved solids). Note different scales for each histogram.

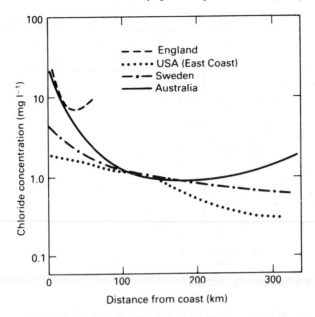

Fig. 2.9 Decrease in chloride content of rainwater with increasing distance inland from the coast (after Junge and Werby, 1958).

Anthropogenic Inputs to the Atmosphere

Carbon Dioxide

In spite of the enormous mass of the atmosphere, the addition of some man-made components has produced small but important changes in the atmospheric concentrations of several of the less abundant species, e.g. CO_2 and SO_2. These extra additions to the atmospheric reservoir are not normally matched by corresponding increases in rates of removal and thus the steady state, which it is assumed existed before human intervention, is destroyed. The box model approach can be used to estimate the proportions of man-made additions which occur in each different reservoir, with the object of assessing long-term dangers. More is known about the behaviour of carbon dioxide than about many other pollutants, and there is much concern at present about the influence of the extra CO_2 on global climate.

First, it is necessary to understand the natural behaviour of CO_2 in the atmosphere and the evidence for changes induced by human activity. Carbon dioxide is a product from the combustion of all

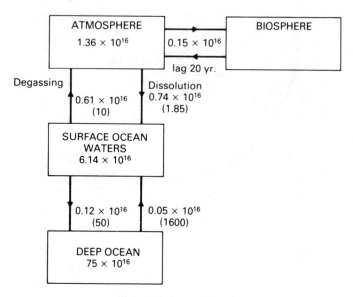

Fig. 2.10   The carbon dioxide cycle. Reservoir units in moles, fluxes in mol yr^{-1}. The residence time of carbon dioxide in each reservoir with respect to the output flux is shown in brackets beneath the magnitude of that flux (from Machta, 1972).

fossil fuels and is discharged directly to the atmosphere. Fig. 2.10 shows the natural reservoirs of carbon which participate in the carbon cycle. On land, carbon dioxide is taken up by vegetation and stored in plants and soil organic matter in the biosphere. The residence time of carbon in the biosphere is 20 years which corresponds to the average lifetime of plants plus the time taken for microbiological decomposition to CO_2 in the soil system. The CO_2 derived originally from the atmosphere is, therefore, returned to it on average after a 20 year lag period. The ocean reservoir is divided into two parts: an upper layer which is well-mixed and in contact with the atmosphere and a deeper, larger reservoir of water. The surface layers maintain an approximate equilibrium with the atmosphere by continued dissolution and degassing of CO_2 but are only mixed with the deeper waters in polar regions where the surface layers, normally warmer and lighter, become colder and heavier than the deep waters and mixing occurs. Elsewhere mixing between the upper and lower layers is poor. The rapid mixing between the atmosphere and surface waters, as compared to the poor mixing between surface and deep layers, is illustrated by the residence times of CO_2 in the two reservoirs. The residence time of CO_2 in the atmosphere with respect to its loss by dissolution into the ocean is 1.85 years, whereas the residence time in the surface layer with respect to transfer into the deeper layers is 50 years.

The carbon dioxide content of the atmosphere was first measured in the early part of the nineteenth century and found to be seasonally variable within the northern and southern hemispheres, due to abstraction by growing vegetation in the spring. In the last two decades a series of measurements at locations remote from local sources of pollution have however clearly demonstrated increasing levels of CO_2 due to pollution (Fig. 2.11). The upward trend is on average 0.8 ppm (parts per million) per year and the mean value of approximately 316 ppm in 1958 had risen to 332 ppm by 1976. These compare with an estimated value of approximately 292 ppm for the nineteenth century. There is also an annual fluctuation showing CO_2 removal when plant growth is at a maximum which, for Mauna Loa (Hawaii), is in the northern hemisphere springtime. Fig. 2.11 also shows the rate at which the concentration of atmospheric CO_2 would have increased if all man-made CO_2 remained in the atmosphere. The observed increase is roughly half that expected. From the carbon cycle it is reasonable to conclude that the missing CO_2 is being transferred to the surface layers of the ocean, since their rates of exchange with the atmosphere are quite rapid. The problem is that the surface layers of the ocean have a capacity only 5–6 times larger than that of the atmosphere and their equilibration with the deeper layers is slow. The following quantitative example shows this.

The residence times given in the CO_2 cycle are computed with respect to a single output and show the influence of that output on the whole reservoir. Where the residence times are low, rates of transfer out of the reservoir are rapid. Therefore, if the residence time of CO_2 in the atmosphere with respect to its flux into the surface layers is 1.85 years then 1/1.85 or 0.54 gives the fraction of that reservoir which can be transferred by that flux in one year. Thus the reciprocal of residence time is a rate constant.

Now consider transfers from the atmosphere to the surface layer and thence to the deeper layers. The biosphere can be ignored since, although it may increase in size in response to increasing CO_2 concentrations, the effect is minor and only introduces a small delay factor into changes in atmospheric composition. The rate constant for CO_2 transfer from the atmosphere to surface layers is 0.54 yr^{-1} and from surface layers to deeper layers is 0.02 yr^{-1}. Ignoring transfers in the reverse direction and assuming that an incremental increase, i, in atmospheric composition is divided amongst the reservoirs in the ratio of their current masses (atmosphere 1.36 $\times 10^{16}$ moles, surface layers 6.14×10^{16} moles), the equilibrium

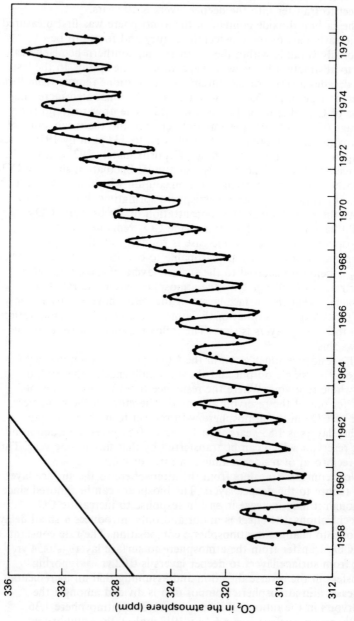

Fig. 2.11 The increase in carbon dioxide concentrations from the burning of fossil fuels. Mean monthly CO_2 concentrations at Mauna Loa (Hawaii). Straight line shows annual rate of increase if all CO_2 remained in the atmosphere (from Woodwell, 1978).

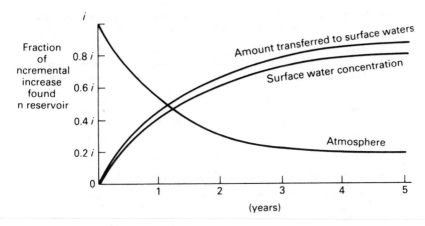

Fig. 2.12   Variations in the reservoir concentrations of carbon dioxide following an instantaneous injection of carbon dioxide into the atmosphere. Amounts expressed as a fraction of the incremental increase *i*.

concentration of the excess CO_2 in the surface layer should initially be given by:

$$\text{Equilibrium concentration} = \frac{i \times 6.14}{1.36 + 6.14} = 0.84i$$

A fraction 0.54 of this amount will transfer in one year ($0.54 \times 0.84\,i$ = $0.45\,i$) and a fraction 0.54 of the remainder ($0.54 \times (0.84 - 0.45)\,i$) in the second year. Fig. 2.12 shows that more than 90% of the equilibrium value is reached after a four-year period from the injection of CO_2 into the atmosphere. During this period transfer to the deeper layers is also occurring. And, because the mass of the deeper layers is so much greater (75×10^{16} moles), a fraction

$\dfrac{75}{75 + 6.14}$ or 0.92 of any increase in the surface layer should represent

the equilibrium concentration in the deeper layers. Table 2.3 shows the equilibrium values in the deeper layers increasing as the surface layer concentrations increase, the two being related by the fraction 0.92. However, although the equilibrium values indicate that a large proportion of the incremental addition ultimately reaches the deeper layers of the ocean, the rate at which this happens is slow. Only a fraction 0.02 of the difference between the actual and equilibrium levels is transferred in one year and the CO_2 levels of the deeper waters increase very much more slowly than do the surface layers. This model is much simplified, since it ignores all reverse transfers and complications caused by the dissociation of CO_2 in water.

Table 2.3   Changes in the CO_2 content of the atmosphere and ocean reservoirs following an instantaneous addition of CO_2 to the atmosphere

Time (yr)	Amount transferred to surface layer*	Corresponding equilibrium level in deep waters*	Cumulative transfer to deep waters*	Surface layer concentration*
1	0.45	0.41	0.008	0.442
2	0.66	0.61	0.020	0.640
3	0.75	0.69	0.034	0.716
4	0.80	0.74	0.049	0.751
5	0.82	0.75	0.063	0.756

* Amounts expressed as a fraction of the original incremental addition, i

Note that at the end of 5 years 75.6 % of the added CO_2 is present in the surface layers of the ocean, 6.3 % in the deep waters and 18.1 % remains in the atmosphere

However, it does predict that additions of man-made CO_2 will be found mainly in the atmosphere and surface layers of the ocean and that relatively little will be transferred to the deeper layers over a short time period. Predictions based on measurements of atmospheric CO_2 levels and estimates of rates of CO_2 input from fossil fuel burning, indicate that the true split is close to 50:50, rather than the estimate of 18:76 obtained from the simplified approach above.

The changes which might result from additions to these two reservoirs must be assessed. Most attention has been focussed on changes in the atsmospheric reservoir, because of the climatic implications. The temperature of the earth is maintained by solar radiation in a wavelength band centred on the visible region. Approximately 19 % of the incoming radiation is absorbed by water vapour, ozone and carbon dioxide. Outgoing radiation at the appropriate wavelengths is also absorbed and reradiated both upwards and downwards. The importance of CO_2 lies in its ability to absorb at the longer wavelengths, so it has relatively little impact on the short wave incoming radiation. However, heat reflected from the earth's surface lies at longer wavelengths than the incoming solar radiation, because the earth's surface temperature is cooler. This reflected heat is absorbed by CO_2 and partly reradiated back to the surface. The main effect of increasing carbon dioxide concentrations, due to fossil fuel burning, is that the gas which is radiating heat to outer space is found at a higher level in the atmosphere, where the temperature is cooler, so that the CO_2 reradiates less effectively. Since less heat is reradiated to outer space the earth's atmosphere must become warmer. This process is called the greenhouse effect, although this is partially a misnomer in that the glass on a greenhouse prevents convection and wind cooling as well as reradiation.

It is estimated that an increase in mean global temperature of 0.5 – 1°C could result from the increased CO_2 levels expected by the end of the century, but it is not clear whether these changes are at least partially offset by increased evaporation rates which would produce more water vapour, and hence cloud cover. Additional cloud would reflect incoming solar radiation and tend to lower the temperature. This type of effect is called negative feedback, where peturbation of a steady state system produces an effect which acts, through a different mechanism, to restore the system to its original steady state condition, or even beyond. Much of the difficulty in assessing the impact of CO_2 pollution on climate is concerned with estimating the relative magnitudes of the different positive and negative feedback processes, since downward variations in the average surface temperature as small as 2–3°C may trigger drastic changes in climate, on the scale of an ice age. Similarly variations of only a fraction of a degree may seriously increase the incidence of severe winters or frost and thus affect the production of sensitive crops. By contrast upward variations may lead to partial melting of the polar ice caps and it has been estimated that natural climatic variations in the past gave sea-level changes of 150 m from glacial maxima to minima. Changes of this magnitude would be sufficient to inundate almost all the fertile coastal areas of the continents.

Sulphur Dioxide

The impact of CO_2 emission to the atmosphere has global implications in terms of climate modification, but other pollutants from combustion processes, especially sulphur dioxide, may have serious local effects. In mass terms sulphur dioxide constitutes the second major addition to the atmospheric reservoir. It mainly arises through the combustion of fossil fuels which often contain a few per cent sulphur. In coal, which has the highest sulphur content of common fuels, this may occur as the mineral iron pyrites FeS_2, which is oxidised during high temperature combustion:

$$2FeS_2 + 3\tfrac{1}{2}O_2 \rightarrow Fe_2O_3 + 2SO_2$$

iron pyrites   iron oxide

A reconsideration of the equilibria discussed on page 27 shows that the dissolution of SO_2 in water gives a more acidic solution than does CO_2, i.e. pH 4.9, as compared to 5.6 for a typical atmospheric partial pressure of CO_2. Rainfall becomes more acid either locally, where meteorological conditions prevent dispersion of SO_2 from polluting areas, or downwind from centres of pollution. For example, Scandinavia lies in the path of winds which disperse SO_2 emission from industrial centres in Europe and acid rainfall results (page 60).

Sulphur dioxide also has major local effects on health. In high concentrations sulphur dioxide is a poison, and in the trace concentrations in which it occurs in polluted cities it slows down the ciliary beat of the lungs which inhibits man's ability to eliminate soot particles. The same meteorological conditions which act to confine SO_2 and prevent its dispersion also retain soot particles and, therefore, the two pollutants are closely correlated. The presence of either pollutant separately would be less damaging, for example soot particles alone would largely be cleared from the lungs by the normal ciliary beat, but together their effect is enhanced. The consequences of this enhancement are illustrated in Fig. 2.13 which shows the increased deaths from bronchial ailments particularly amongst the aged, during some of the great London smogs of 1952 (Fig. 2.14). There is a clear correlation with the occurrence of soot

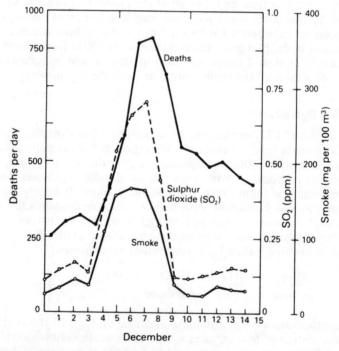

Fig. 2.13 Daily air pollution and deaths in London, December 1952. The synergistic relationship between smoke and SO_2 causes increased death rates, especially amongst the elderly (from Wilkins, 1954).

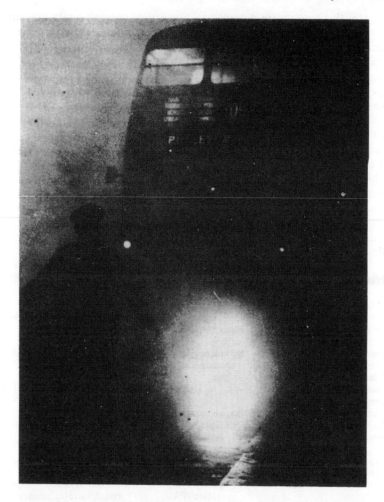

Fig. 2.14 The London smog of 1952. This marked an important turning point in the control of air pollution in the United Kingdom (courtesy of Keystone Press Agency).

and sulphur dioxide. The type of effect where one pollutant reinforces another is called synergism.

These two air pollutants, soot and sulphur dioxide, have been substantially reduced in London over the last 20 years, primarily as a result of economic factors which dictated the switch from coal to oil, but partly as a result of legislation to control domestic coal burning. However, a different type of smog has become increasingly prevalent in other cities, especially Los Angeles. The first measures taken to control this smog were designed to break the synergistic relationship between soot and SO_2 by controlling dust emissions. Home incinerators and the burning of public refuse was prohibited and limits were placed on dust emissions from industry. Considerable improvement in visibility resulted but the typical smog symptoms of eye irritation and plant damage remained. A research programme was initiated to discover the origin and nature of the atmospheric pollutants and it soon became apparent that there were important differences between Los Angeles and London smog. For example, London smog is strongly reducing due to SO_2 but Los Angeles smog is often strongly oxidising. Other significant differences emerged which indicated that industrial and domestic sources were not mainly responsible (Table 2.4) but which pointed to an automobile source, exacerbated by the city's unusual geographical and meteorological situation. Los Angeles lies within a bowl created by surrounding hills. Temperature inversion effects, which confine pollutants to the air above the city (Fig. 2.15), are common.

Table 2.4 Comparison of Los Angeles and London smog

Characteristic	Los Angeles	London
Air temperature	24 to 32°C	−1 to 4°C
Relative humidity	< 70%	85% (+fog)
Type of temperature inversion	Subsidence, at a few thousand metres	Radiation (near ground) at a few hundred metres
Wind speed	< 3 ms^{-1}	Calm
Visibility	< 0.8 to 1.6 km	< 30 m
Months of most frequent occurrence	August–September	December–January
Major fuels	Petroleum	Coal and petroleum products
Principal constituents	O_3, NO, NO_2, CO, organic matter	Particulate matter, CO, S compounds
Type of chemical reaction	Oxidative	Reductive
Time of maximum occurrence	Mid-day	Early morning
Principal health effects	Temporary eye irritation (PAN)	Bronchial irritation coughing (SO_2/smoke)
Materials damaged	Rubber cracked (O_3)	Iron, concrete corroded

Pollutants from automobiles, however, are not directly responsible for the smog. The primary pollutants emitted by cars follow a daily cycle with a bimodal form (Fig. 2.16) arising from high traffic density in the mornings and evenings. However, the main effect of Los Angeles smog is apparent at mid-day, lying between the peaks in traffic density. The reason for this is that the primary pollutants emitted by cars are not directly responsible, but they react to produce secondary pollutants and this causes smog. Therefore, the

Fig. 2.15 Photochemical smog in Los Angeles (courtesy of South Coast Air Quality Management District).

peak in secondary pollutants, i.e. the smog itself, follows the peak of primary pollutants. Fig. 2.16 also shows that there is no peak in secondary pollutants to follow the evening rush hour, which indicates that the reactions which produce Los Angeles smog require the presence of sunlight, i.e. they are photochemical reactions.

The primary pollutants emitted by a car engine are carbon monoxide, hydrocarbons and nitrogen oxides, the latter being the principal precursors of smog. Concentrations of the nitrogen oxides, NO and NO_2, in the atmosphere are maintained by the reaction cycle shown in Fig. 2.17. This situation is normally stable and maintains ozone levels at a uniformly low value. However, if organic materials, e.g. unburnt or partially burnt fuels, are added to the system, then oxygen atoms are used to decompose the organic molecules into reactive fragments, known as free radicals. These fragments oxidise nitric oxide very efficiently, so that the

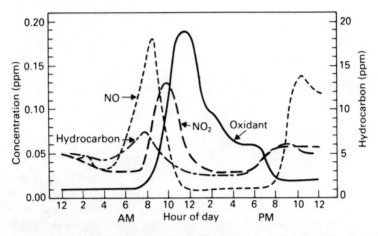

Fig. 2.16 Daily cycle of primary and secondary pollutants in smog. The peak in secondary pollutants (oxidants) lags behind the primary pollutants (nitrogen oxides, hydrocarbons) emitted by automobiles (from the Continuous Air Monitoring Program in Cincinnati 1962–63, US Dept of Health, Education and Welfare (1965)).

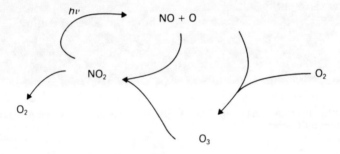

Fig. 2.17 Schematic representation of the pseudo-equilibrium state between nitrogen oxides, oxygen and ozone in the atmosphere.

concentration of NO decreases. Ozone is now no longer used up in reactions with NO and its concentration rises, eventually reacting with the organic molecules to speed up the overall production of free radicals. These can react with a variety of atmospheric components to produce the aldehydes and ketones which characterise Los Angeles smog. Reaction with NO_2 is responsible for the eye irritant peroxy acyl nitrate (PAN), i.e. $CH_3 COOONO_2$. The relationships between the major species which interact to produce Los Angeles smog are shown in Fig. 2.16 as a function of time. The daily cycle is clearly evident. Emission of primary pollutants from cars is followed by photochemical conversion to oxidants which subsequently produce a variety of products from hydrocarbons.

3 The Crust

The crust is the largest reservoir in the earth-air-water factory (mass $= 2.2 \times 10^{22}$ kg) and is also chemically the most heterogeneous. Not only do all the stable elements in the periodic table occur in the crust, but the form of occurrence varies widely. Oxygen, the most abundant element, occurs as a major structural component in no fewer than 1300 classes of minerals. By contrast some rare second and third row transition elements, e.g. rhenium, rhodium and osmium, appear to form only one naturally-occurring compound, although trace amounts of these elements may substitute into common minerals in place of more abundant elements. The number of minerals formed by an element depends partly on its abundance in the crust and partly on its chemical properties. Each particular mineral forms, and is stable, only within a certain range of environmental conditions (temperature, pressure, pH, redox potential[†] etc.). The huge variety of different minerals found on the earth's surface arises because there is often a large overlap in the range of conditions under which different minerals are stable and because minerals may persist for considerable periods of time in unstable conditions, where their rates of reaction are slow.

The role of water is critical. Rock minerals formed deep in the crust under high temperatures and pressures show little tendency to react when uplifted to the surface and exposed in an arid environment. By contrast the same minerals exposed in a continuously humid climate rapidly lose their fresh surfaces and become extensively weathered. Everyday experience provides one example of the importance of water in chemical reactions. The oxidation of metallic iron to iron oxide occurs with the liberation of energy, i.e. spontaneously, under surface conditions:

$$4Fe + 3O_2 \rightarrow 2Fe_2O_3 \qquad\qquad \Delta G^{\ominus} = -1484.5 \text{ kJ}$$

The reaction proceeds very slowly in a totally dry atmosphere but is dramatically accelerated in the presence of moisture. The iron oxide Fe_2O_3 is the red-brown rust which forms on exposed ironwork in the presence of water. Attempts to prevent rust formation are mainly associated with preventing the access of moisture, either by keeping metallic objects clean and dry or by coating them with a film of material which is more resistant to oxidation, e.g. chromium plating, or water-repellant, e.g. silicone waxes. These simple preventive measures exploit the fact that the kinetics, or rate of reaction, of iron and oxygen are slow if water is absent.

Water can exert an important influence on most chemical reactions, including those involving rocks and minerals. Its ubiquitous presence in the crust is emphasised by the representation of this part of the earth-air-water factory as a liquid extractor plant, in which crushed rock debris is continuously leached by water.

The Role of Water

Water plays a variety of interrelated roles in crustal processes and these can be considered principally as those of a solvent, carrier and catalyst. Firstly water is a solvent. In the liquid state a molecule of H_2O has the shape of an isoceles triangle, where the spacing between the oxygen and hydrogen atoms is 0.096 nm compared to 0.151 nm between the two hydrogen atoms. Oxygen is more electronegative[†] than hydrogen and consequently the bonding electrons are drawn towards the oxygen atom. The resulting separation of charge creates a dipole, with the hydrogen end of the molecule having a slight positive charge and the oxygen end a slight negative charge. Since oxygen is among the most electronegative of elements the dipole moment of water is high, almost twice as large as for its sulphur analogue H_2S. The strongly polar nature of water is the reason for its outstanding properties as a solvent. The force of attraction between the dipole and ions on a crystal surface is responsible for the destruction of crystals in solution, the process called solution or hydrolysis. Clearly polar solvents are most effective in dissolving crystals where there is charge separation between the components ions, thus water readily breaks down ionic bonded crystals but is much less effective in dissolving most covalent compounds.

The second important role of water is as a medium for transportation, and again the influence of the dipole moment is critical. Upon release from a crystal lattice, ions are immediately surrounded by an envelope of water molecules because the charge on the surface of the ion creates an electrostatic potential gradient, which is millions of volts per centimetre, near the surface of the ion, and which attracts the polar water molecules. The subsequent behaviour of the dissolved ion is determined by whether the water molecules are attracted more strongly to other water molecules than to the ion, i.e. whether the ion will be hydrated or whether it will be forced from the solution, i.e. precipitated. The strength of the ion-water attractive force for ions of the like charge varies according to the inverse square of the ionic radius. Larger ions are less hydrated, simply because the same electronic charge is distributed over a larger surface area. The relative degree of hydration of ions is expressed by the parameter Z/r, where Z is the electronic charge and r is the ionic radius. This quantity is called the ionic potential. Fig. 3.1 is a plot of ionic charge versus ionic radius for many common ions. Lines have been drawn connecting elements in the same rows of the periodic table. The division of the plot into three areas is made according to the different behaviour of ions with respect to water, as revealed by the ionic potential. Water acts as an important medium for the transportation of ions with an ionic potential less than 3 (strongly cationic) or greater than 12 (strong complex anions). When the ionic potential lies between 3 and 12, ions are precipitated as hydroxides and are effectively immobile in aqueous environments.

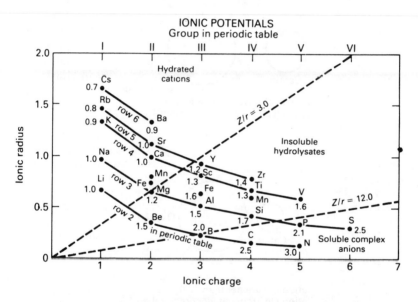

Fig. 3.1 Relationship between ionic charge and ionic radius. Lines connect elements in the same row of the periodic table. Values of the ionic potential Z/r distinguish elements which behave as hydrated cations, insoluble hydroxides and soluble complex ions.

The mechanism by which ionic potential controls mobility and transportation is illustrated in Fig. 3.2. When the ionic potential is less than 3, the electrostatic field at the surface of the cation is sufficient for bonding with water dipoles. However smaller ions, with higher ionic potentials, possess a sufficiently large electrostatic field to cause the electrons in the outer shell of the oxygen atom to be

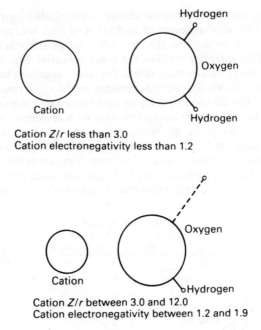

Cation Z/r less than 3.0
Cation electronegativity less than 1.2

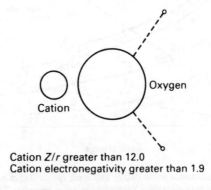

Cation Z/r between 3.0 and 12.0
Cation electronegativity between 1.2 and 1.9

Cation Z/r greater than 12.0
Cation electronegativity greater than 1.9

Fig. 3.2 The relationship between cation size, cation-oxygen separation and the force of repulsion between a cation and the hydrogen ions belonging to a water molecule.

pulled nearer to the cation. This weakens the bond between oxygen and hydrogen in the water molecule to the extent that the binding forces between oxygen and one of the hydrogens can be overcome. A hydrogen ion is therefore ejected into solution and an insoluble hydroxide is formed. The extreme case of this type of interaction is found for ions with an ionic potential greater than 12, where the force of attraction between the cation and the oxygen atom is sufficiently great to expel both hydrogens from the water molecule. A soluble oxy-anion results and the H^+ in solution causes an increase in solution acidity. The effect of ionic potential is very important in controlling the extent to which various ions released from crystal breakdown are mobile and can be transported.

The third important role played by water in the crust is that of a catalyst, acting to speed up chemical reactions without itself undergoing any change. Sometimes catalysts act by reducing the energy barriers to chemical reactions through the formation of an intermediate or activated compound (page 22). An alternative, and probably more important, way of viewing the catalytic role of water is through the collision theory of reaction rates, which suggests that chemical reactions only occur when the reactant molecules collide with approximately the correct orientation and have sufficient internal energy to react. Water acts as a catalyst by allowing ions to approach each other closely without spatial constraint, therefore encouraging collisions with the correct orientation.

The roles of water as a solvent, carrier and catalyst are very evident in the weathering reactions which dominate this section of the earth-air-water factory. As in any other chemical reaction, the products of weathering depend on the nature of the reactants and the reaction conditions. An elementary knowledge of mineral structures is necessary before weathering processes can be understood.

The Structure of Silicate Minerals

Chemical compounds between elements which readily form ions have structures which mainly reflect the relative dimensions of the component positive and negative ions, since this controls how closely the two ions may approach each other. Therefore, where bonding is almost entirely electrostatic, i.e. ionic bonding†, crystal structure may be deduced by envisaging the ions as close-packed spheres with unlike charges in contact and like charges separated. The problem is purely geometric; spheres of different sizes must be arranged to give the densest possible packing. Table 3.1 gives the ionic radii of some common elements and also shows the percentage ionic character for the bond between each ion and oxygen. Note that although there are two major types of bond, ionic or covalent†, the bonding for virtually all compounds is to some extent intermediate between these two types.

Where bonds are predominantly ionic the structure of a compound can be predicted from the radius ratio rule. For the

Table 3.1   Ionic radii of selected elements and the percentage ionic character of their bonds with oxygen.

Element		Ionic radii* (nm)	% Ionic character
Aluminium	Al^{3+}	0.051	60
Barium	Ba^{2+}	0.034	84
Calcium	Ca^{2+}	0.099	79
Carbon	C^{4+}	0.015	23
Iron	Fe^{2+}	0.074	54
	Fe^{3+}	0.064	69
Magnesium	Mg^{2+}	0.066	71
Manganese	Mn^{2+}	0.080	72
Potassium	K^{+}	0.133	87
Silicon	Si^{4+}	0.042	48
Sodium	Na^{+}	0.097	83
Strontium	Sr^{2+}	0.112	82

* Ionic radii for elements in octahedral coordination.

bonding between an element and oxygen we define the radius ratio as:

$$\text{Radius ratio} = \frac{\text{Ionic radius of ion}}{\text{Radius of } O^{2-}} = \frac{r_k}{r_{oxygen}}$$

If the radius ratio lies between 1 and 0.73 the pattern of closest packing is achieved if each ion is surrounded by eight ions of the opposite charge, as its closest neighbours. Radius ratios between 0.73 and 0.41 reach closest packing if each ion is surrounded by six ions of opposite charge and for values of $0.41 - 0.22$ each ion can only have four nearest neighbours. Radius ratio values are given in Table 3.2 and it can be seen that predicted number of nearest neighbours (the coordination number) agrees well with that observed in crystal structures. The radius ratio rule is successful in predicting the structures of mainly ionic compounds, even to the extent that ions which lie on the borderline between two categories are seen to exhibit either coordination number.

However, this simple model is of less value in predicting the structures of compounds in which the bonding is largely covalent. The metals cadmium and calcium have similar ionic radii (Ca^{2+} 0.099 nm, Cd^{2+} 0.097 nm) and although the radius ratio rule correctly predicts the structures of their compounds with oxygen, it fails conspicuously for the more covalent sulphide compounds. The bonds of aluminiun and silicon with oxygen dominate most silicate minerals and have an almost equally ionic and covalent character and the coordination numbers of aluminium and silicon in silicates correspond to that predicted by the radius ratio rule. However, it would be totally wrong to conclude from this that

Table 3.2 Radius ratio values

Critical radius ratio	Predicted coordination	Ion	Radius ratio r_k/r_{oxygen}	Commonly observed coordination numbers
	3	C^{4+}	0.16	3
	3	B^{3+}	0.16	3, 4
0.225				
	4	Be^{2+}	0.25	4
	4	Si^{4+}	0.30	4
	4	Al^{3+}	0.36	4, 6
0.414				
	6	Fe^{3+}	0.46	6
	6	Mg^{2+}	0.47	6
	6	Li^+	0.49	6
	6	Fe^{2+}	0.53	6
	6	Na^+	0.69	6, 8
	6	Ca^{2+}	0.71	6, 8
0.732				
	8	Sr^{2+}	0.80	8
	8	K^+	0.95	8-12
	8	Ba^{2+}	0.96	8-12
1.000				
	12	Cs^+	1.19	12

electrostatic forces are responsible for the overall structure. In fact, the large degree of covalent character exerts an important influence on the behaviour of silicate minerals during weathering.

The fundamental structural unit of all silicate minerals is the silicon tetrahedron which consists of a central silicon atom fitting almost exactly into the space between four closely-packed oxygen atoms, arranged spatially such that they occupy the four corners of a tetrahedron. The relative sizes and positions of silicon and oxygen in a single tetrahedron are accurately portrayed in Fig. 3.3, but unfortunately this method of representation lacks clarity for more complex silicate structures. Thus individual tetrahedra are usually drawn as shown in Fig. 3.4, where the distances between adjacent atoms has been exaggerated relative to the ionic size. This approach has been adopted in the present text.

Since silicon has a valency of four and oxygen is divalent, each tetrahedron carries a net 4− charge. The bonding within, and between, silica tetrahedra is almost equally ionic and covalent, but the bonds between metal ions and silica tetrahedra are largely ionic. Therefore, it would be expected that water, as a polar solvent, would more readily remove metal ions from silicates than break down the bonds between tetrahedra.

The simplest of the true silicate minerals is composed of isolated

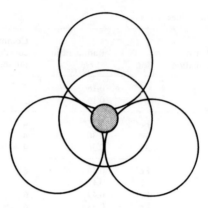

Fig. 3.3 Silicon and oxygen close packing in the silicon tetrahedron. The shaded silicon atom lies below the central oxygen atom, but above the three oxygen whose centres lie in a single plane.

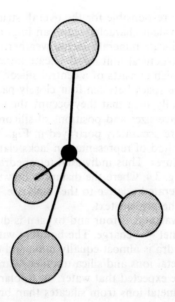

Fig. 3.4 The silicon tetrahedron with bond lengths exaggerated.

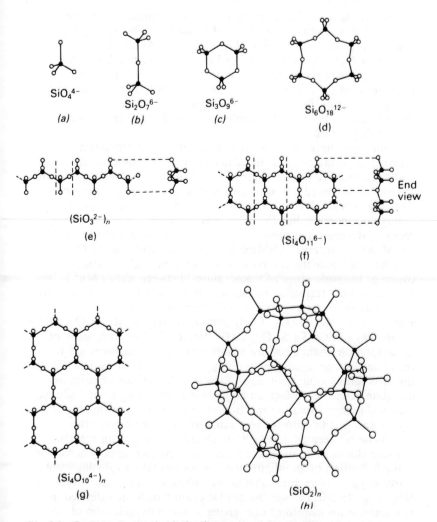

SiO_4^{4-}

(a)

$Si_2O_7^{6-}$

(b)

$Si_3O_9^{6-}$

(c)

$Si_6O_{18}^{12-}$

(d)

$(SiO_3^{2-})_n$

(e)

$(Si_4O_{11}^{6-})$

(f)

End view

$(Si_4O_{10}^{4-})_n$

(g)

$(SiO_2)_n$

(h)

Fig. 3.5 Common structural units in silicate minerals: a, neosilicates; b, c and d, sorosilicates; e and f, inosilicates; g, phyllosilicates; h, tektosilicates.

tetrahedra of $(SiO_4)^{4-}$ units (Fig. 3.5a), surrounded by cations which maintain a charge balance. Such minerals are called neosilicates of which the olivines Mg_2SiO_4 (forsterite) and Fe_2SiO_4 (fayalite) are examples. The structure of these minerals can be conceived as one of isolated tetrahedra surrounded on each of their four sides, and linked to adjacent tetrahedra, by metal cations.

Another class of silicates (sorosilicates) is composed of multiple groups of linked tetrahedra. Between 2 and 6 tetrahedra may be linked together in short chains (Fig. 3.5b, c, d). Akermanite $(Ca_2MgSi_2O_7)$ is a sorosilicate in which two silica tetrahedra are linked together by sharing a common oxygen atom (Fig. 3.5b). Linkages between adjacent tetrahedra need not terminate but can be continuous, as in the cyclosilicates. In the mineral beryl $(Be_3Al_2Si_6O_{18})$ the silica tetrahedra are joined together to form a six-membered ring (Fig. 3.5c).

In another, more complex class, silica tetrahedra are linked together to form linear chains of indefinite extent (inosilicates). Two kinds of chain are found: single chains with a silicon:oxygen ratio of 1:3 and cross-linked double chains with a silicon:oxygen ratio of 4:11 (Fig. 3.5e, f). The single chain structure occurs in an important family of minerals called the pyroxenes, which are present in many rocks of the crust and upper mantle. Simple members of this family are $MgSiO_3$ (enstatite), $CaMgSi_2O_6$ (diopside) and $NaAlSi_2O_6$ (jadeite). However the composition of these minerals is variable, owing to the wide range of isomorphism[†] between Ca^{2+}, Mg^{2+}, Fe_i^{2+} and other transition metals. The cross-linked double chains are present in an important class of rock-forming minerals called amphiboles e.g. $Mg_7Si_8O_{22}(OH)_2$ (anthophyllite), $Ca_2Mg_5Si_8O_{22}(OH)_2$ (actinolite) and $Ca_2Fe_5Si_8O_{22}(OH)_2$ (tremolite). The complexity of the amphibole chain structure means that the cations necessary to maintain electroneutrality may occupy a variety of sites which are not structurally equivalent, and so the possibilities for isomorphism are quite extensive. In fact, an amphibole crystallising from a silicate melt will accept a small proportion of any available cation, in one site or another. In both the pyroxenes and the amphiboles the cations lie between adjacent silicate chains and link them together.

From chains of an indefinite extent the next step in polymerisation is the formation of an infinite two-dimensional sheet (phyllosilicates). Three oxygens of each tetrahedra are linked with adjacent tetrahedra (Fig. 3.5g). In effect this is the double chain inosilicate extended in two dimensions instead of one, giving a silicon:oxygen ratio of 2:5. Common phyllosilicate minerals include $KAl_2(AlSi_3O_{10})(OH)_2$ (muscovite), $Mg_3(Si_4O_{10})(OH)_4$ (talc) and $Mg_6(Si_4O_{10})(OH)_8$ (serpentine) and all clay minerals. These minerals form layer or sandwich structures and usually have a planar shape which reflects their internal structure. In some clay and serpentine minerals the sheets are rolled into tubes (Fig. 3.6).

Finally, three-dimensional networks of silicate tetrahedra are known as the tektosilicates. The simplest possible tektosilicate is quartz (SiO_2), in which the silica tetrahedra are linked together to make a three-dimensional network (Fig. 3.5h) by having each oxygen

(a)

(b)

Fig. 3.6 Electron micrographs of phyllosilicates: a, face and edge views of kaolinite stacks, showing smooth surfaces of plates separated by incipient cleavage. Picture width 17 μm; b, Tube-like structure of halloysite, a member of the kaolinite family. Picture width 2 μm (courtesy of N. K. Tovey).

atom shared between adjacent tetrahedra to give a silicon:oxygen ratio of 1:2. The structure of the feldspars is analogous to that of quartz, except that some of the silicon sites are occupied by aluminium. Since aluminium is trivalent the excess valencies of the tetrahedral framework are balanced by other metals such as Na^+, K^+ and Ca^{2+}. For example, in the mineral $KAlSi_3O_8$ (orthoclase) three out of every four tetrahedral sites are occupied by silicon and one by aluminium, with charge balance maintained by having one K^+ present in the space available in the network for each Al present in the tetrahedral framework. The structure of the sodium feldspar (albite) is the same, but the calcium feldspar (anorthite) has a higher aluminium:silicon ration $(CaAl_2Si_2O_8)$ which necessitates the presence of a divalent ion to achieve charge balance.

Weathering Processes

Weathering processes in the earth-air-water factory are represented by the liquid extractor plant (Fig. 1.5). Here rock materials undergo a slow reaction utilising carbon dioxide from the atmosphere and water from the hydrosphere. The gas-liquid interactions (page 24) therefore provide an important basis for the solid-liquid interactions which are the central topic of this section.

The products of weathering reactions may be either soluble or insoluble. The soluble materials, together with colloidal and fine-grained particles, are removed from the weathering site and are ultimately transported to the oceans, whilst certain of the solid residues remain in the soil system as a prior stage in a much slower journey to the ocean basins. The separation of these two components takes place in what is essentially a liquid extractor plant, where large volumes of water are percolated through a permeable mass of rock fragments. The removal of the dissolved components and their influence on river chemistry will be discussed later in the chapter. The solid residues which result from rock weathering reactions are discussed here.

The conversion of rock into soil is a striking metamorphosis. Rocks, which are absolutely dead, chemically inactive, relatively hard and non-porous, are converted into soils which are soft, porous, chemically active and very much alive. This change in character results from exposure to surface conditions. Most of the rocks in the earth's crust have been formed at high temperatures and pressures, often in the absence of air or water, or both. On the earth's surface, temperatures and pressures are very much lower and air or water, or both, are present. Weathering is the adjustment of rocks and minerals to these new and different conditions.

In the weathering zone the massive structure of rock is broken down to smaller particles by physical weathering and the original minerals changed by chemical weathering to different forms which are more stable under surface conditions. Usually chemical and physical weathering processes operate concurrently, but for the sake of simplicity each may be considered separately.

Physical weathering takes place initially as a result of unloading. When the confining pressures of the earth's crust are lessened by uplift and erosion of the overlying material, the release of internal stress by expansion causes the development of cracks and joints. Additional strains can now be imposed by a variety of processes which include thermal expansion and contraction, wetting and drying, abrasion by transported rock fragments and, most importantly, by crystal formation, particularly the formation of ice which expands on crystallisation to exert a prising action within a crack or joint. These physical weathering mechanisms are important because they increase the total surface area of the rock material exposed to air and water which are the agents of chemical weathering (Fig. 3.7).

Mechanisms of Chemical Weathering

It is possible to distinguish four different mechanisms involved in

Fig. 3.7 Physical weathering. Intense frost action in the Dry Valleys, Antarctica has produced characteristically sharp fractures in this boulder (courtesy of R. Raiswell).

chemical weathering: dissolution, oxidation, hydrolysis and acid hydrolysis. Various combinations of these four processes, or even all four together, may occur simultaneously during the breakdown of some rock minerals.

Dissolution   The simplest of all weathering reactions is the solution of soluble minerals. Because of its polar nature, water is most effective in dissolving ionic solids, such as NaCl (halite).

$$NaCl\,(s) \xrightleftharpoons{H_2O} Na^+\,(aq) + Cl^-\,(aq)$$

Note that the chemical reaction for the solution of halite does not contain H^+, so that this process would be expected to be independent of pH.

Oxidation   Free oxygen is important in the decay of all rocks containing reduced substances, especially iron and sulphur. At surface temperatures, oxidation reactions are slow and a variety of oxidisable materials, e.g. wood, coal and petroleum, can remain in contact with oxygen in the air until quite high temperatures before oxidation, i.e. combustion, occurs. Water also speeds up the oxidation of minerals as it does with metals (page 45), probably by the dissolution of minute amounts of material. In these cases water fulfils a catalytic role. The oxidation of reduced Fe(II) in silicates such as Fe_2SiO_4 forms colloidal $Fe(OH)_3$ which, on dehydration, gives a variety of stable Fe(III) oxides.

$$Fe_2SiO_4\,(s) + 2\tfrac{1}{2}O_2\,(g) + 5H_2O\,(l) \rightarrow 2Fe(OH)_3\,(s) + H_4SiO_4\,(aq)$$
$$Fe(II) \phantom{+ 2\tfrac{1}{2}O_2\,(g) + 5H_2O\,(l) \rightarrow 2}Fe(III)$$

All the iron oxides formed in this manner contain iron (III) and are conspicuous because of their bright colours. The colour of the simple oxide Fe_2O_3 (haematite) is characteristically dusky red, whereas the hydrated oxides FeOOH (goethite, lepidocrocite) are often yellow to brown. It is usual to represent these hydrated oxides as $Fe(OH)_3$, which actually approximates to the composition of the amorphous iron (III) oxide precipitated in the laboratory by adding alkali to Fe^{3+} solutions. Such laboratory precipitates are, however, believed to undergo an aging process, involving dehydration and recrystallisation, which gives haematite either with or without prior conversion to goethite. The common occurrence of these products is a reflection on the great stability and insolubility of iron (III) oxides.

Oxidation is also indirectly important as a weathering process through the action of living organisms, which use atmospheric oxygen to metabolise organic matter to carbon dioxide. The carbon dioxide so produced is the source of most of the protons consumed in weathering reactions. In biologically productive soils the amount of CO_2 from this source in the soil waters may be up to one hundred times the amount expected from equilibrium with atmospheric CO_2.

$$\text{Organic } C + O_2 \rightarrow CO_2$$
$$CO_2 + H_2O \rightarrow H_2CO_3 \rightarrow H^+ + HCO_3^-$$

In many cases organic matter is not completely degraded to CO_2 and the partial breakdown products may have acidic properties. Such organic acids possess carboxyl (COOH) or phenolic (OH) groups which can dissociate (page 75) to supply an additional source of protons in the soil zone.

$$RCOOH \rightleftharpoons RCOO^- + H^+ \left.\begin{array}{l} \\ \end{array}\right\}$$
$$R\,\text{OH} \rightleftharpoons R\,\text{O}^- + H^+ \quad\Big\}\; R = \text{large organic entity}$$

Hydrolysis The process of hydrolysis similar to that of dissolution except that the water, as well as dissolving part or all of the mineral being weathered, also reacts chemically with the resulting ions. Two examples are shown below, using the simple igneous mineral Mg_2SiO_4 (forsterite) and the important sedimentary mineral $CaCO_3$ (calcite).

$$Mg_2SiO_4 (s) + 4H_2O (l) \rightleftharpoons 2Mg(OH)_2(aq) + H_4SiO_4(aq)$$
$$\Updownarrow \quad \text{Silicic acid}$$
$$2Mg^{2+} (aq) + 4OH^- (aq)$$
$$CaCO_3 (s) + 2H_2O (l) \rightleftharpoons Ca(OH)_2(aq) + H_2CO_3 (aq)$$
$$\Updownarrow \quad \text{Carbonic acid}$$
$$Ca^{2+} (aq) + 2OH^- (aq)$$

Silicic acid is a weak acid (page 116) and is virtually undissociated at pH < 9.9, i.e. in most surface environments. Similarly carbonic acid is a weak acid and, therefore, the solutions resulting from these hydrolysis reactions are alkaline because both $Mg(OH)_2$ and $Ca(OH)_2$ are moderately strong alkalis. Since weak acids are virtually undissociated, the significant dissociation of the alkalis leads to an excess of OH^- over H^+ in the solution. Therefore, the pH is greater than 7.

Similar results are obtained with other minerals, since most rock minerals yield a reasonably strong alkali and a very weak acid on hydrolysis. It should be noted that the example of dissolution of soluble salts, above, can be regarded as a hydrolysis reaction in which the resulting acid and alkali (HCl and NaOH) are equally strong. The reaction then yields equal amounts of H^+ and OH^-, which effectively neutralise each other in solution.

There are minerals for which the hydrolysis reaction yields an acid solution:

$$Fe_2(SO_4)_3 \text{ (s)} + 6H_2O \text{ (l)} \rightleftharpoons Fe(OH)_3 \text{ (s)} + 3H_2SO_4 \text{ (aq)}$$

$$\Updownarrow$$

$$6H^+ \text{ (aq)} + 3SO_4^{2-} \text{ (aq)}$$

Iron (III) hydroxide is a very weak alkali and sulphuric acid is a strong acid. In certain situations (e.g. mine workings), Fe^{3+} and SO_4^{2-} ions are derived from the oxidation of FeS_2, and the mine waters may have pH values as low as 1 or 2. In general however minerals yielding a weak alkali and a strong acid on hydrolysis are uncommon at the earth's surface.

Acid hydrolysis   Acid hydrolysis is identical to ordinary hydrolysis except that the water performing the hydrolysis contains dissolved acids which make it substantially more effective as a weathering agent. The acidity comes from a variety of sources, of which the dissolution of atmospheric CO_2 and SO_2 to give H_2CO_3, H_2SO_3 and H_2SO_4 have been discussed (pages 25–28). Additional acids, including H_2CO_3 and organic acids, are supplied by oxidation.

For convenience, in the acid hydrolysis reactions below the source of the acidity is taken as H_2CO_3 and the same starting minerals are used as in the ordinary hydrolysis reactions.

$$Mg_2SiO_4 \text{ (s)} + 4H_2CO_3 \text{ (aq)} \rightleftharpoons 2Mg^{2+} \text{ (aq)} + 4HCO_3^- \text{ (aq)}$$
$$CaCO_3 \text{ (s)} + H_2CO_3 \text{ (aq)} \rightleftharpoons Ca^{2+} \text{ (aq)} + 2HCO_3^- \text{ (aq)}$$

As with ordinary hydrolysis the solutions resulting from acid hydrolysis reactions are alkaline since the Mg^{2+} (or Ca^{2+}) and HCO_3^- ions are hydrolysed in solution to form the weak acid H_2CO_3 and the relatively strong alkali $Mg(OH)_2$ (or $Ca(OH)_2$). Thus one result of acid hydrolysis is to neutralise any acid contained in the water performing the weathering. This is an important finding since man's consumption of fossil fuels leads to elevated levels of CO_2 and SO_2 in the atmosphere and also in rainwater through the dissolution processes discussed on page 27. In most cases the extra acidity given to rainwater by anthropogenic activities is neutralised during weathering of soil minerals. However where the soil cover is very thin and non-calcareous (page 78), as in parts of Scandinavia, the rainwater acidity may not be completely neutralised before the water enters rivers and lakes. The addition of hydrogen ions and the consequent decrease in pH may have important economic implications. The newly hatched fry of salmon are very sensitive to acidity, more so than trout for example, and even a relatively small decrease in natural pH to 4 or 5 may jeopardise fishing and fish-rearing activities in such waters. Salmon and trout have been eliminated from many Scandinavian rivers and lakes, respectively, probably due to increased acidity resulting from the emission, oxidation and long-distance transport of air pollutants such as sulphur dioxide.

So far a distinction has been drawn between ordinary and acid hydrolysis, yet this division is somewhat unrealistic since most waters involved in weathering contain some acidity. There is, therefore, a spectrum of hydrolysis processes extending from pure hydrolysis at neutral pH to intense acid hydrolysis in waters with a low pH. The most important factor in determining to which part of this spectrum a particular environment belongs will obviously be water acidity the sources of which have been discussed earlier.

Silicate Weathering Mechanisms

The example chosen to illustrate acid hydrolysis of an igneous rock mineral (forsterite) is a particularly straightforward example since the forsterite dissolves completely. With other minerals the situation is often more complicated and the reaction products will be various dissolved species plus an aluminosilicate residue. Despite the huge variety and complexity of silicate minerals in the crust, it is possible to represent their weathering reactions schematically. This approach has the advantage that attention is sharply focused on the many points of similarity in the weathering of different silicate minerals, which may be normally obscured by the complexity of the reaction equations.

Fig. 3.8 shows the termination of a silicate lattice at the crystal surface. Note that residual charges are created by bond breakage at fracture surfaces because only the complete 3-dimensional network is electrically neutral. These residual charges are the focus for attack by protons during acid hydrolysis. This reaction can be envisaged as an ion exchange, involving the loss of a metal ion from the lattice to be replaced by protons. The products are a metal-deficient hydrated silicate plus an aqueous solution of metal bicarbonate. Removal of the metal ions leaves a protonated silicate lattice whose Si—O—Si bonds may be severed by hydrolysis (Fig. 3.8). Continued hydrolysis progressively nibbles away at the edges of the silicate lattice resulting in the complete hydration and mobilisation of silicon as $Si(OH)_4$. Not all the silicate lattice may be broken down and the residual solid material is a metal-deficient hydrated silicate (a clay mineral) with unique and important properties.

Silicate weathering can be summarised, as below, where Me^{2+} represents a metal ion, in this case divalent.

$$\text{Silicate} \diagdown^{O}_{O} \diagup \text{Si} \diagup^{O^-}_{O} \diagup \text{Me}^{2+} \text{ (s)} + 2H^+HCO_3^- \text{ (aq)} \rightleftharpoons$$

$$\text{Silicate} \diagup^{O}_{O} \diagdown \text{Si} \diagdown^{O^-H^+}_{O^-H^+} \text{ (s)} + \text{Me}^{2+} \text{ (aq)} + 2HCO_3^- \text{ (aq)}$$

$$\text{Silicate} \diagdown^{O}_{O} \diagup \text{Si} \diagdown^{O^-H^+}_{O^-H^+} \text{ (s)} + 2H_2O \text{ (l)} \rightleftharpoons$$

Cation deficient silicate (s) + $Si(OH)_4$ (aq) + $2OH^-$ (aq)

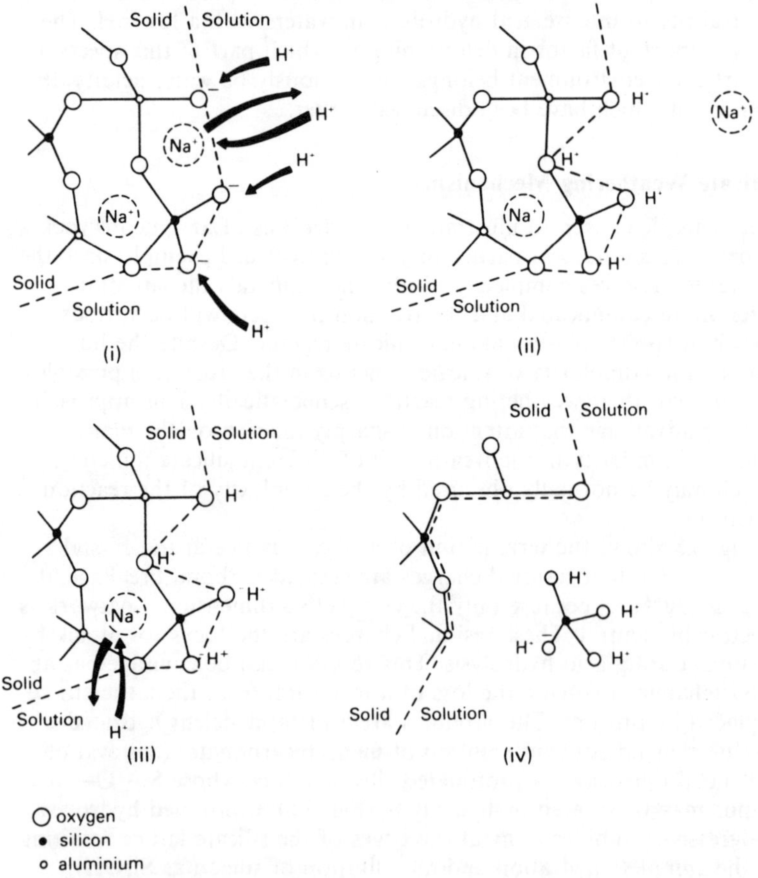

○ oxygen
● silicon
○ aluminium

Fig. 3.8 Weathering reactions at the surface of a feldspar: (i) Broken bonds become protonated and ionic-bonded Na^+ is exchanged with H^+ from solution; (ii) Protonated lattice; (iii) Further exchange of Na^+ for H^+ causes complete protonation of the edge tetrahedron; (iv) Edge tetrahedron completely removed to solution as H_4SiO_4.

Fig. 3.9 Chemical weathering. Residual corestones of fresh and crumbly granite isolated in a clay matrix produced by chemical weathering (from Ollier, 1962).

The products of weathering are, therefore, a solid silicate which is partially degraded and hydrated, dissolved silica and a metal bicarbonate solution. In the case of forsterite all the starting material will be dissolved ultimately, but this is not typical (Fig. 3.9). Other silicates behave as in the schematic weathering reaction above.

The Controls on Weathering Reactions

With even this limited knowledge of silicate structures and the mechanisms of weathering reactions, it is now possible to understand which conditions will promote rapid weathering. Observations of stable, deeply weathered soil profiles in tropical regions suggest that the critical variables which control the rate of weathering are temperature, water flow rate and water composition. The effect of temperature can be easily seen. The higher the temperature, the faster the rate of chemical reaction. In general an increase in temperature of $10°C$ approximately doubles the rate of reaction, so that silicate weathering in the tropics, where the mean annual temperature is $20°C$, proceeds about 50% faster than in the temperate zone, where the mean annual temperature is $12°C$.

The effect of water flow rate can be understood by a simple

model. Reactions which readily proceed to completion, i.e. in which the equilibrium point lies well over towards the reaction products, are the exception in the natural environment and in many acid solution or hydrolysis reactions the equilibrium lies towards the starting materials. For example, the weathering of sodium feldspar to the clay mineral kaolinite plus soluble components proceeds as follows:

$$2NaAlSi_3O_8(s) \quad + 9H_2O(1) + 2H^+HCO_3(aq) \rightleftharpoons$$
Sodium feldspar

$$Al_2Si_2O_5(OH)_4(s) + 2Na^+(aq) + 2HCO_3^-(aq) + 4H_4SiO_4(aq)$$
Kaolinite

(3.1)

The equilibrium constant for this reaction can be derived from free energy values.

$$\log K_E = \frac{-\Delta G^\ominus}{5.707}$$

The free energy values for the species contained in Equation (3.1) are shown in Table 3.3. Using these values it can be shown that the overall free energy change for this reaction is $+69.1$ kJ. and the equilibrium constant is $10^{-12.1}$. Pure solid phases are assumed to have a concentration of one and, therefore,

$$K_E = \frac{[Na^+]^2[HCO_3^-]^2[H_4SiO_4]^4}{[H_2CO_3]^2}$$

Clearly the equilibrium point for this reaction is far over to the left. Therefore, the forward reaction only predominates until the water in

Table 3.3   Free energy values for the species in Equation (3.1) (from Krauskopf, 1967)

Species		Free energy of formation, $\Delta G^\ominus$
Formula	Name	(kJ mol^{-1})
$NaAlSi_3O_8$ (s)	Sodium feldspar	-3698.7
$Al_2Si_2O_5(OH)_4$ (s)	Kaolinite	-3776.9
$H_2O(1)$	Water	-237.2
H_2CO_3 (aq)	Carbonic acid	-623.4
H_4SiO_4 (aq)	Silicic acid	-1308.8
Na^+ (aq)	Sodium ion	-261.9
HCO_3^- (aq)	Bicarbonate ion	-587.0

contact with the feldspar has the concentrations of HCO_3^- and Na^+ predicted from the equilibrium constant, at which point the rates of backward and forward reactions become equal and no further weathering occurs. If the water in contact with the fieldspar is flushed away and replaced by pure water the equilibrium is disturbed and Le Chatelier's principle[†] predicts that the forward reaction will again predominate until the equilibrium concentrations of HCO_3^- and Na^+ are reached. The equilibrium constant derived above for the weathering of sodium feldspar to kaolinite is of the same order as that for the weathering of other silicate minerals. In these circumstances, where the equilibrium point lies well towards the reactant side, the reaction can only be induced to go to completion if the reaction products are constantly, flushed away by the arrival of new reactants. Water flow rate, therefore, exercises an important control on rates of weathering.

The importance of water flow rate in flushing away even sparingly soluble weathering products can be emphasised by comparing soils developing in the humid tropics with those in arid environments. The rapid water flow rates in humid weathering environments can eventually result in the removal of all but the most insoluble aluminium and iron oxides. By contrast in arid environments there is no substantial dissolution and even very soluble chloride and sulphate-bearing minerals commonly accumulate close to the surface.

The effect of water composition on rates of weathering, as distinct from the nature of the final weathered product, operates essentially through the hydrogen ion concentration, $[H^+]$, of the solution. In weathering processes, the higher the concentration of H^+ the more rapid the attack on the basic components of the silicate crystal lattice. The hydrogen ion concentration also affects the relative solubilities of some products of weathering. At the normal range of pH in the weathering zone (pH 5–9), silicon is more soluble than aluminium. If a pH value can be held in this range whilst the preferential removal of silica proceeds to completion, the resulting soil will be a mixture of complex oxides of iron and aluminium, generally called lateritic or ferralitic. This is the situation which has developed at freely-drained sites on stable land surfaces in the tropics. Soils which develop under broadly similar conditions at higher latitudes also contain iron and aluminium oxides but these are associated with quartz. In the extreme case of podzolic soils the upper mineral horizons consist principally of quartz, and both iron and aluminium oxides have been leached away and segregated in deeper horizons (Fig. 3.10). In many cases, however, the lower rates of reaction, coupled with the short time available since the deposition of fresh glacial sediments (10 000–250 000 years), have not permitted the full development of podzolisation in most temperate

Fig. 3.10 Podzolisation. Iron and aluminium oxides have been leached from the upper soil horizon leaving a pale residual quartz layer (courtesy of H. S. Gibbs).

areas. Nevertheless, the simple thermodynamic treatment developed earlier can be used to analyse the environmental conditions under which ferralitic and podzolic soils develop, following an approach used by Curtis (1970).

In natural waters there are two main forms of dissolved aluminium: the simple ion Al^{3+} and the complex ion $Al(OH)_4^-$. The former predominates under very acid conditions and the latter in neutral and alkaline environments. By writing aluminium oxides as $Al(OH)_3$, the contributions made by each ion to the solubility of total aluminium can be assessed.

$$Al(OH)_3 \rightleftharpoons Al^{3+} + 3OH^- \tag{3.2}$$

$$Al(OH)_3 + H_2O \rightleftharpoons Al(OH)_4^- + H^+ \tag{3.3}$$

From the free energies of formation for each of these species the overall free energy change in each reaction can be determined, and the equilibrium constant derived. The free energies of formation of $Al(OH)_3$, Al^{3+}, $Al(OH)_4^-$ and OH^- are -1144.3, -481.1, -1302.5 and -157.3 kJ mol^{-1}, respectively. So the free energy change in reaction (3.2) is

$$\Delta G^\circ = 3(-157.3) + (-481.1) - (-1144.3)$$
$$\Delta G^\circ = +191.3 \text{ kJ mol}^{-1}$$
$$\log K_{(3.2)} = \frac{-\Delta G^\circ}{5.707} = \frac{-191.3}{5.707} = -33.5$$
$$\therefore \quad [Al^{3+}][OH^-]^3 = 10^{-33.5}$$

A similar series of steps gives $K_{(3.3)}$:

$$[H^+][Al(OH)_4^-] = 10^{-13.9}$$

Since $[H^+][OH^-] = 10^{-14}$ we can put $[OH^-] = \dfrac{10^{-14}}{[H^+]}$

So $\dfrac{[Al^{3+}]}{[H^+]^3} = 10^{8.5}$

As before $[H^+][Al(OH)_4^-] = 10^{-13.9}$

Each of these expressions represents the equations of lines drawn on a set of axes describing concentration of dissolved aluminium and pH (Fig. 3.11). For pH values lower than that represented by the intersection of the two lines (pH = 5.6), Al^{3+} predominates over $Al(OH)_4^-$ and the contribution made by the latter to the solubility of aluminium can be ignored. The reverse applies to pH values higher than the intersection point.

The curve formed by the two lines represents values of total dissolved aluminium for which there is equilibrium between solid aluminium oxides and dissolved aluminium. Within the curve solid forms of aluminium are stable, outside the curve dissolved forms are stable. The horizontal line represents concentrations of amorphous silica in equilibrium with dissolved silica, and this line cuts the aluminium curve at approximately pH 4.5. This composite diagram displaying the solubilities of aluminium and silica as a function of pH can now be used to distinguish the conditions under which ferralitic and podzolic soils develop.

In the shaded area of Figure 3.11, aluminium, as Al^{3+}, is more soluble than silica and will consequently be preferentially removed from a soil profile. The solubility relationships predict that aluminium remains more soluble until the pH rises above 4.5. It can, therefore, be predicted that podzols will only occur where pH values are below 4.5. For values in excess of 4.5 quartz is more soluble than

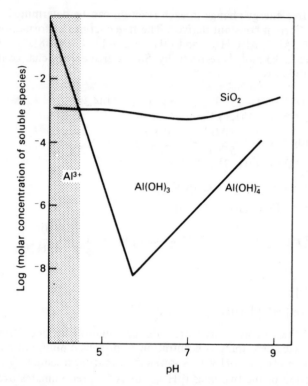

Fig. 3.11 The solubility of aluminium and silicon as a function of pH. In the shaded area aluminium is more soluble than silicon and may be preferentially removed from a weathering profile.

aluminium oxides and can be preferentially removed from the weathering zone. Ferralitic soils should, therefore, be restricted to pH values in excess of 4.5. The correlation of theoretical and natural occurrences is close. Measurements show that the bleached upper horizons of podzolic soils have pH values of 3.5 – 5.0, whilst ferralitic soils lie in the range of 5.0 to 6.0. The products of weathering are, therefore, closely controlled by the hydrogen ion concentration of the weathering solutions.

It cah also be predicted from Fig. 3.11 that, inside the aluminium curve and above the silica line, both aluminium oxides and silica would be immobile and substantially retained within the weathering profile. Under these circumstances the silica and aluminium oxides may combine together with cations to form secondary minerals known as clays. In this way chemical weathering provides the raw materials necessary for the synthesis of clay minerals.

Clay Minerals: The Solid Products of Weathering Reactions

Structural Properties

Many important differences between clay minerals and their primary rock mineral precursors are a function of their structure. Clay minerals are built up from two structural units. One is a sheet of silica tetrahedra arranged as a hexagonal network in which the tips of the tetrahedra all point in the same direction (Fig. 3.12). This is the same unit described earlier for the phyllosilicates. The other structural unit consists of two layers of closely packed oxygen or hydroxyl groups in which Al, Fe or Mg atoms are embedded so that each is equidistant from six oxygens or hydroxyls. Most clay minerals can be thought of as sandwiches of the two rudimentary structural units: the tetrahedral sheet and the octahedral sheet. The simplest type of sandwich is made up of a single layer of silica tetrahedra with an aluminium octahedral layer on top (Fig. 3.13).

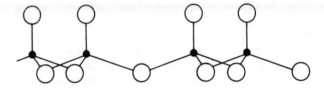

Fig. 3.12   A sheet of silica tetrahedra.

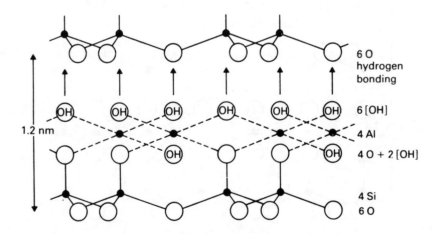

Fig. 3.13   The structure of a 1 : 1 clay mineral (kaolinite).

This is the basic structure of the kaolinite family of minerals, which are sometimes called 1:1 minerals because of the condensation of one tetrahedral sheet with one octahedral sheet, so that the tips of the tetrahedra and one of the hydroxyl layers of the octahedra form a common layer. In kaolinite, the type member of this group of minerals, the flat double sheets, are stacked one above the other in packages which are held firmly together by hydrogen bonding[†] between the upper hydroxyl layer of the octahedral sheet and the basal oxygens of the tetrahedral layer. Even though this type of bonding is weak it has the important property of holding successive packages of the two layers closely together and consequently discouraging the entry of any foreign ions.

The other main type of sandwich is the 2:1 structure, consisting of an octahedral filling between two tetrahedral slices. These three layers are condensed together with the tips of the silica tetrahedra pointing inwards, each forming part of the octahedral sheet (Fig. 3.14). In smectite minerals the octahedral sites may be occupied by magnesium, iron or other small metal ions as well as by aluminium. Aluminium may also occur in some of the tetrahedral sites in place of silicon. These isomorphous substitutions of divalent metals for trivalent aluminium and of trivalent aluminium for tetravalent silicon leave the clay structure with a surplus negative charge. This charge is balanced by a variety of interlayer cations, principally Al^{3+}, Ca^{2+}, Mg^{2+}, K^+, Na^+ and H^+. It should be noted that there can be no

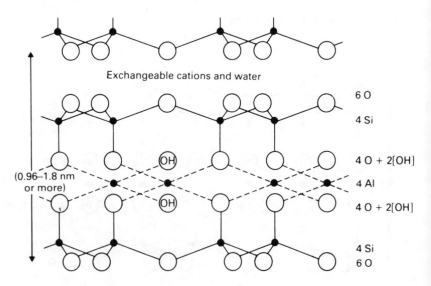

Exchangeable cations and water

6 O
4 Si

4 O + 2[OH]
4 Al
4 O + 2[OH]

4 Si
6 O

(0.96–1.8 nm or more)

Fig. 3.14   The structure of a 2:1 clay mineral (smectite).

hydrogen bonding between individual 2:1 packages because oxygens at the base of one layer of tetrahedra are faced by oxygens, rather than hydroxyls. Therefore, the individual packages are not held closely together. Water and other polar solvents can penetrate into this position with the result that the lattice of smectite clays expands with uptake of water, and shrinks on drying.

Mica minerals are phyllosilicates rather than clay minerals, but their structure is considered here because of their close relationship to illite. Micas are structurally similar to smectites with the distinction that there is a very regular substitution of aluminium for silicon in the tetrahedral layer, with only slight variations from a 1:6 ratio. The surplus negative charge created is balanced by K^+ ions which fit neatly into holes in the base of the tetrahedral sheet (Fig. 3.15) and hold the 2:1 units firmly together in stacks. Primary mica minerals often weather to form illites which differ from micas in having some isomorphous substitution in the octahedral sheet, less substitution of aluminium for silicon in the tetrahedral layer and consequently less K^+ in the interlayer position. There is also some replacement of K^+ by exchangeable cations. Illites are, therefore, poorly crystalline and less well-ordered than micas.

Clay minerals, therefore, share many common structural features. In particular they are based on Si–O and Al–O bonds which are to a large degree covalent and not susceptible to rupture by water, which is a polar solvent. The bonding between Mg–O, (and all the other

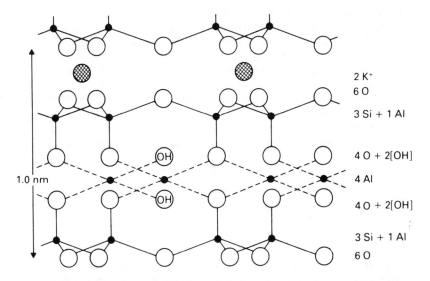

Fig. 3.15 The structure of muscovite.

cations), is more ionic (Table 3.1) and the metal ions are more readily removed in solution. Minerals with a high degree of isomorphous substitution are, therefore, less stable under conditions of intense hydrolysis.

Formation by Weathering

The clay mineral that is formed in any particular weathering environment, or indeed whether any clay mineral is formed, depends upon the provision of raw materials by weathering and the intensity of leaching, the latter primarily controlled by water flow rate and pH. Since the pH of most weathering environments is greater than 4.5, Si can be removed in preference to Al in aluminosilicates and the Si:Al ratio provides a crude measure of leaching intensity. Smectites have a high ratio of Si to Al and a high contents of bases, both as interlayer cations and as components of the octahedral sheet. Therefore, they will only accumulate where there is a low intensity of leaching and weathering of basic rocks to provide the necessary raw materials. Under more intense leaching cations and silica are preferentially removed. The predominant clay mineral is then likely to be kaolinite, which has a 1:1 Si:Al ratio and no cations. Under the most extreme leaching the end products of weathering will be gibbsite $Al(OH)_3$ and goethite $FeOOH$.

The intensity of leaching is a function of rainfall and drainage. On stable land surfaces in the humid tropics, where weathering and leaching are intense and have continued over a very long period, freely-drained sites are mantled in gibbsitic and kaolinite clays, regardless of parent rock, and smectites are found in poorly-drained depressional sites (Fig. 3.16). In Hawaii, which is entirely built up of basic lavas, the predominant clay mineral in the soils changes from smectite to gibbsite over a range of rainfall from 450–4500 mm yr^{-1} (Fig. 3.17). As the leaching intensity increases the silicate clays become progressively more unstable. A general weathering sequence can be drawn up for the different clay minerals (Fig. 3.18). In temperate regions weathering has operated at a lower intensity for a shorter

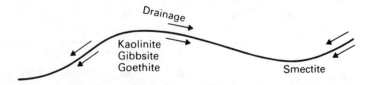

Fig. 3.16 The influence of drainage and relief on clay mineralogy. The relatively rapid water flow rates associated with high relief result in the preferential removal of cations and silica.

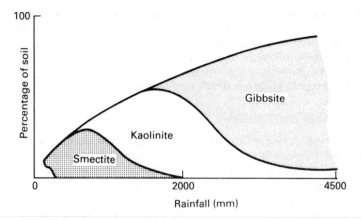

Fig. 3.17 The influence of climate on clay mineralogy in Hawaii. The relatively rapid water flow rates associated with high rainfall result in the preferential removal of cations and silica (after Sherman, 1952).

time. Repeated glaciations have stripped the weathered mantle from vast areas and redistributed the material as till, loess and fluvial deposits. The clay mineralogy of these recent clay deposits is consequently very varied.

The weathering reactions by which clay minerals are derived from

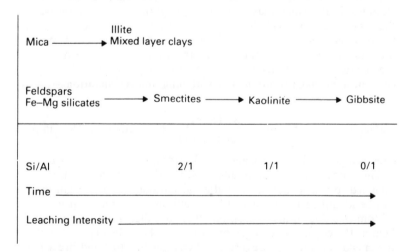

Fig. 3.18 The production of clay minerals by weathering as a function of leaching intensity.

rock silicates are an important step in the chemical separations which take place within the liquid extractor plant, yet there are other more subtle solid-liquid interactions which are fundamental to plant life and, therefore, indirectly to all living systems. The structural changes which accompany the transformation from rock silicate to clay mineral produce materials which can undergo ion exchange.

Ion Exchange and Fertilisers

Ion exchange is the ability of minerals, or synthetic chemicals, to hold ions temporarily on their surfaces, where they are resistant to leaching but from which they may be replaced by other ions. In general, the electrostatic forces holding the ions are sufficiently weak to allow their easy replacement. Therefore, if particles containing one type of adsorbed ion are added to an electrolyte solution containing different ions, some of the adsorbed ions will be released to the solution and replaced by ions from solution. This process operates constructively in the cycling of plant nutrients and in the application of fertilisers to the soil, but is probably more familiar in its use for softening water. Hard water contains Ca^{2+} ions derived from the weathering of $CaCO_3$-bearing rocks. The large amounts of dissolved Ca^{2+} form an insoluble scum of calcium stearate with the anionic (stearate) part of the soap molecule, which cannot form a lather until the Ca^{2+} has been removed. Therefore, in hard-water areas much more soap, or sodium stearate, is required for washing processes. If hard water containing Ca^{2+} is passed over crystals of a zeolite (an aluminosilicate which may be natural or synthetic) containing Na^+, then the Ca^{2+} displaces the Na^+ in the crystal. The water gains the harmless Na^+ ion and loses Ca^{2+} and is said to be softened. When most of the Na^+ has been displaced from the zeolite it must be regenerated, which is accomplished by passing through a concentrated NaCl solution. The softening and regeneration reactions are as follows:

$$\text{Na-zeolite(s)} + Ca^{2+}(aq) \quad \underset{\text{Regeneration}}{\overset{\text{Softening}}{\rightleftharpoons}} \quad \text{Ca-zeolite(s)} + 2Na^+(aq)$$

This is a particularly clear and simple example of ion exchange.

The exchangeability of an adsorbed ion depends on how it is attached to the crystal. Several alternative attachment mechanisms can be identified in the environment, although in practice it is seldom possible to distinguish one from another. Firstly, ions may be held at the corners and edges of particles, where the exposed parts of the crystal structure possess residual charges due to bond breakage. In the crystal lattices of silicate minerals, electroneutrality is maintained by the balance between electronegative and electropositive atoms. At the edge of the crystal this balance cannot operate locally, since the fractured surface will in some cases sever bonds to create centres of surface negative charge, from partly uncoordinated oxygens, or positive charge, from partly uncoordinated silicon or metal atoms. The surface negative charges are balanced by adsorbed cations and the positive charges by adsorbed anions. The quantity of ions held in this way by coarse particles is infinitesimal but, for a given weight of material, the total surface area increases dramatically as the particle size decreases. Clay particles are very small, usually $< 2 \mu$m diameter, and therefore have a large surface area and a significant proportion of imbalanced charges at the crystal edge.

Alternatively, many clay minerals possess a permanent surface negative charge due to the isomorphous substitution of *trivalent* aluminium for *tetravalent* silicon in the tetrahedral layer and also substitution of *divalent* metal ions for *trivalent* aluminiums in the octahedral layer. Both substitutions result in a deficiency of positive charge on the clay mineral lattice which is balanced by cations adsorbed at the interlayer positions. Isomorphous substitution occurs to a greater degree in smectites than in illites, for example, and the charge deficiency, and exchange capacity, are therefore greater.

Hydroxyl groups exposed at the edges of a crystal may dissociate and produce an additional surface negative charge to be balanced by cations from solution.

$$\text{Mineral} - O - H(s) + Me^+ (aq) \rightleftharpoons \text{Mineral} - O - Me(s) + 2H^+(aq)$$

In this form the dissociation can clearly be seen to be pH-dependent. At low pH the exchange equilibrium will lie to the left side of the above reaction and the exchange capacity of the exposed hydroxyl groups will consequently be low. At high pH the reverse applies and dissociation of H^+ is encouraged, the resulting charge deficiency being balanced by metal ions.

A further source of cation exchange, associated with soil organic matter and quantitatively important in the soil system, is salt formation by organic acids which provides a source of anions (negative charge) which may be balanced by adsorbed metal ions in a similar way as occurred with hydroxyl groups, and is similarly pH dependent. Soil organic acids usually contain the carboxylic group

COOH or phenolic group ⬡ OH, most cation exchange being

associated with the former. A consideration of the dissociation constants of the two types of acid shows why. Acetic acid, a typical carboxylic acid, has a dissociation constant K_D of 1.75×10^{-5}.

$$CH_3COOH \rightleftharpoons H^+ + CH_3COO^-$$

$$K_D = \frac{[H^+][CH_3COO^-]}{[CH_3COOH]} = 1.75 \times 10^{-5}$$

Hence for $[H^+] > 1.75 \times 10^{-5}$, the ratio $\frac{[CH_3COO^-]}{[CH_3COOH]} < 1$ and the acid is essentially undissociated. Conversely for $[H^+] < 1.75 \times 10^{-5}$ the acid is dissociated and the anion is available to hold soil cations. Comparison of a wide range of carboxylic acids shows that values of K_D usually lie between 10^{-3} and 10^{-5}, clearly indicating that these acids are dissociated where soil pH is greater than 5. By contrast, phenolic groups are often only weakly acidic ($K_D = 10^{-7}$ to 10^{-10}) and may not be appreciably dissociated in many soil systems.

The ions adsorbed on clay and organic surfaces can be replaced by other cations from the soil solution according to the law of mass action. For example, a clay particle has a surface negative charge which will be balanced by adsorbed cations. These cations are in equilibrium with dissolved cations in the soil water. Should an ion, such as potassium, be extracted from the soil water by plant roots, more potassium will be released into solution by the clay in order to restore equilibrium (Fig. 3.19). If potassium ions are added to the soil water, e.g. by the weathering of potassium feldspar or by the addition of a soluble potassium fertiliser such as KCl, the K^+ in turn displaces other cations from the clay mineral surface until equilibrium is reached. The adsorbed cations constitute a nutrient reservoir within the soil system. The cation exchange properties of the soil are of fundamental importance to plant growth because nutrient ions which are adsorbed by clays are no longer soluble in the ordinary sense. This means that they are not directly leached by water draining through the soil, but are held in a form available to plant roots, either through the medium of the intervening soil water or by direct exchange between the root and clay particle.

Potassium, calcium and magnesium are essential plant nutrients

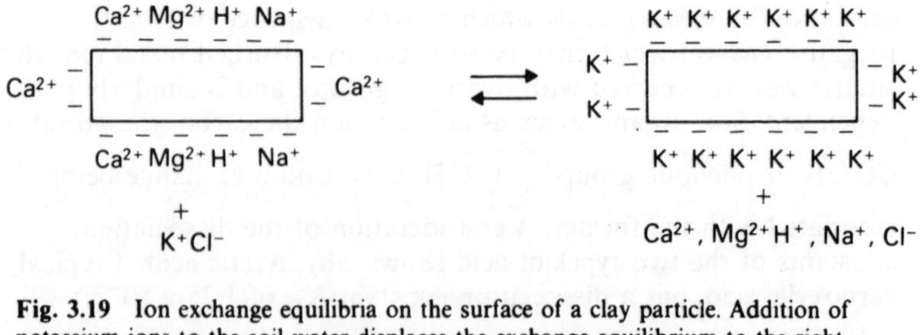

Fig. 3.19 Ion exchange equilibria on the surface of a clay particle. Addition of potassium ions to the soil water displaces the exchange equilibrium to the right. Removal of potassium ions from solution displaces the equilibrium to the left.

which are involved in cation exchange. Nitrogen and phosphorus, on the other hand, may be available to plants only as the anions NO_3^- and PO_4^{3-}, whose behaviour is very different. The anion exchange capacity of soils arises mainly from hydroxyl groups exposed at the edges of clay flakes, which may be replaced by soil anions. Anion exchange capacity is, however, very small relative to cation exchange capacity, and also the nitrate ion is not strongly held. In consequence any nitrate ion released into the soil water, either through the mineralisation of organic nitrogen or as soluble nitrate fertiliser, is rapidly leached into streams or lakes unless quickly utilised by plants. Washout of nitrate from soil is an increasing problem, firstly because intensive systems of agriculture require high inputs of nitrogen fertilisers, and secondly because the subsequent entry of nitrates into streams and lakes can increase nutrient levels and cause eutrophication (page 90). It should be noted that nitrogen applied in the form of ammonium salts is held more firmly by cation exchange in the form of the ammonium ion NH_4^+.

In contrast to nitrate the phosphate ion is strongly held by the soil, to some extent as an adsorbed anion but principally as insoluble iron and aluminium phosphates, at low pH, and calcium phosphates, at high pH. Therefore, acid highly-weathered soils containing large amounts of iron and aluminium oxides and hydroxides fix phosphate very strongly in forms which are largely unavailable to plants. Alkaline soils, containing high concentrations of Ca^{2+}, similarly render phosphate insoluble and less available to plants. Maximum concentrations of utilisable phosphate therefore occur at intermediate pH values between 6 and 7.

The ability of clay minerals to undergo ion exchange is a most important factor in holding many nutrients which are an essential component of soil fertility. It emphasises that weathering is not only a destructive process, but is also constructive in that the clay mineral products possess a significant ion exchange capacity, in contrast to the silicate starting materials.

The Land-based Hydrosphere

The mechanisms of chemical weathering so far considered have been applied to weathering processes in the soil system, the focus of interest being primarily on the solid products of weathering reactions as opposed to the dissolved ones. The major differences between dissolved and solid weathering products lies in the greater mobility of the former. Solid particles may reside in the soil system for many thousands of years whereas dissolved components are more rapidly transported via rivers and lakes to the oceans. The relative proportions of the dissolved and solid products from any weathering environment depend on the minerals present in the bedrock, or more specifically on their solubility. For example, acid hydrolysis of a sedimentary mineral like $CaCO_3$ proceeds much more rapidly than for typical igneous rock minerals and moreover may leave no solid residue (page 59). Because of this freshwaters in drainage basins with carbonate bedrock generally contain much more dissolved material, which is measured as total dissolved solids (TDS) and called hard, as compared to waters draining igneous catchments which are usually soft. Some of the consequences of hard water are discussed on page 74. The chemical differences between waters draining igneous and calcareous terrains are summarised in Table 3.4. The calcareous waters are distinguished by the predominance of Ca^{2+} and HCO_3^- over other dissolved species, high TDS values and low or negligible solid residues from weathering. By contrast the igneous waters are low in TDS and have a more variable composition. These differences are entirely as expected from the simplified treatment of weathering reactions. However, there is a wide variability in composition between different rivers which cannot be explained solely as a result of bedrock composition. The reasons for these differences in river composition have been discussed by Gibbs (1970).

Table 3.4   A comparison between the chemical properties of hard and soft water

Property	Igneous waters	Calcareous
TDS	Low	High
pH	6–8	7–9
Cations	Na^+, K^+, Mg^{2+} etc.	Ca^{2+}, Mg^{2+}
Anions	HCO_3^-, H_4SiO_4	HCO_3^-
Weathered solids	Clay minerals	None

Global Variations in River Chemistry

For every river for which reasonably reliable data is available, the TDS can be plotted against the ratio $\left(\dfrac{Na^+}{Na^+ + Ca^{2+}} \right)$, as shown in Fig. 3.20. The data fall within an envelope which has a boomerang shape and which allows the identification of three mechanisms which control world water chemistry.

Atmospheric Precipitation   Rain-water can be regarded as very dilute sea water since most of its dissolved ions are derived from the remnants of bubble bursting at the sea surface. This means that Na will be the dominant cation, giving a ratio of $Na^+/Na^+ + Ca^{2+}$ close

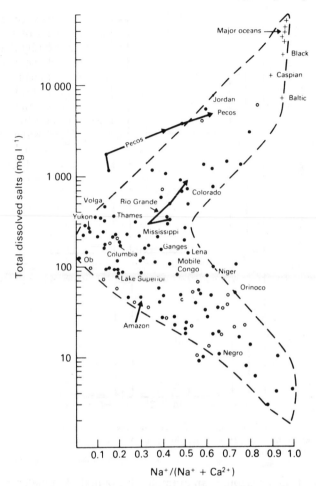

Fig. 3.20 Variations in the weight ratio $Na^+/(Na^+ + Ca^{2+})$ as a function of the total dissolved solids (TDS) content of the world's surface waters (from Gibbs, 1970; copyright 1970 by the American Association for the Advancement of Science).

to 1.0, and, because of the great dilution of sea salt with pure water, atmospheric precipitation will be low in TDS. Therefore, any river whose chemistry is dominated by rain-water input will fall on the lower right of Fig. 3.20.

Typical examples are the tropical rivers of Africa and South America, which have sources in well-weathered, flat terrain where the rainfall is high. Because of this any contribution of weathering to the dissolved load of the river is likely to be small. The Rio Tefé, a

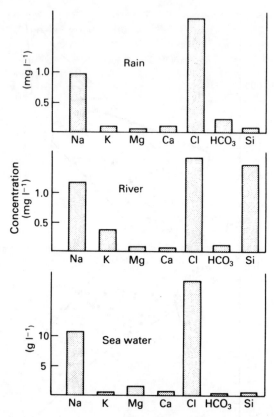

Fig. 3.21   The composition of Rio Tefé river water compared to average rainwater and sea water (from Gibbs, 1970; copyright by the American Association for the Advancement of Science).

tributary of the Amazon, is an example of a rainfall dominated river, and its composition is compared to that of the Tefé basin rainfall and sea water in Fig. 3.21. For all the dissolved ions, except potassium and silicon, the concentrations in rain and river water are remarkably similar, confirming atmospheric precipitation as the dominant mechanism in this river basin. The increased amounts of potassium and silicon are due to some weathering of igneous rock minerals which occurs in tributary streams draining into the Tefé. Not unexpectedly, since one is derived by dilution of the other, the composition of sea water and rain-water collected in the Tefé basin are very similar, although there is an approximately thousand-fold difference in absolute concentration. It is important to note that precipitation dominance is by no means limited to coastal rivers. The

Rio Tefé is approximately 1700 km from the Atlantic Ocean and the plot of Cl$^-$ concentrations of rain-water as a function of distance from the coast (Fig. 2.9) shows that this is in a region where the Cl$^-$ concentration reaches a constant or background level, independent of the distance from the coast. Despite the precipitation influence reaching its minimum, the contribution of salts from this source is more important than from weathering.

Rock Dominance   Rivers controlled by rock dominance are those whose chemistry is dominated by weathering reactions in their drainage basins. They tend to show intermediate values for TDS and occupy approximately the first half of the $Na^+/Na^+ + Ca^{2+}$ axis. The position a river occupies along this axis will depend on the nature of the minerals being weathered in its drainage basin, e.g. rivers draining areas of carbonate rocks tend to appear at the far left of Fig. 3.20, whereas those derived from weathering in igneous basins have higher values for the $Na^+/Na^+ + Ca^{2+}$ ratio.

Although the Amazon is often thought of as a typical tropical river, which should plot near the bottom right or precipitation dominated area of Fig. 3.20, it is a good example of a river whose chemistry is controlled by rock dominance. This is because approximately 85 % of the dissolved material in the Amazon is derived from weathering in the Andes mountains where rock dominance is the main factor.

Rock dominance and atmospheric precipitation have been treated as separate categories, but the two sources are not mutually exclusive and many rivers exhibit properties between the two extreme compositions. Therefore, as Fig. 3.20 shows, it is more realistic to think of rock and precipitation dominances as the two end members of a continuous series.

Evaporation/Crystallisation   The third important mechanism controlling the chemistry of rivers is the evaporation/crystallisation cycle which occurs in some river basins. This process only occurs in hot arid regions where evaporation rates greatly exceed atmospheric precipitation. Rivers for which this process is important fall in the upper right area of Fig. 3.20 forming a continuous series between the end members of a rock dominated river and sea water, the latter plotting at the extreme upper right corner of Fig. 3.20. Rivers in this series are formed by the evaporation of rock dominated water, which increases the TDS. Eventually continued evaporation will precipitate $CaCO_3$ and increase the $Na^+/Na^+ + Ca^{2+}$ ratio of the river water. The increase in TDS and $Na^+/Na^+ + Ca^{2+}$ ratio gives the positive slope in Fig. 3.20 for the rock-dominated sea water series of rivers. Certain rivers, e.g. the Pecos and Rio Grande, show a transition from rock dominance towards the sea water end member as they progress downstream. This trend is indicated by the arrows in Fig. 3.20. Although sea water is the end member of this series, it should be noted that simple evaporation and precipitation of $CaCO_3$ are not sufficient to convert river water into sea water. Other processes must also occur to change the proportions of different dissolved species (see Chapter 4). However, in view of the fact that the cation composition of rain-water is inherited from sea water, it is not surprising to find that sea water and atmospheric precipitation both have the same $Na^+/Na^+ + Ca^{2+}$ ratio, although their TDS values are greatly different. Their similarity in $Na^+/Na^+ + Ca^{2+}$ ratios are responsible for the vertical limit to the right extremity of Fig. 3.20.

So far river composition has been discussed only in terms of cations, but an analogous plot of TDS against $Cl^-/Cl^- + HCO_3^-$ can also be produced. This data, together with the other main ideas developed in this section, is summarised in Fig. 3.22. Mass balance calculations can be used to make quantitative estimates of the relative contributions made by atmospheric precipitation and rock weathering. The results of such a calculation are presented in Table 3.5 for three rivers, each representing an end member from the two mixing series discussed above. Chemical analyses of rain-water and yearly precipitation data have been used to estimate the atmospheric contribution and the difference between the yearly outflow of each ion and its supply by precipitation give the rock contribution. The three main types of river can be clearly distinguished. Variations in relief, vegetation and bedrock composition are all of secondary importance compared to the interplay between atmospheric precipitation, rock dominance and evaporation/crystallisation.

The Chemistry of Lakes

So far this discussion of the land-based hydrosphere has been solely concerned with rivers. Whilst it is true that the chemistry of lakes reflects the chemical composition of the water draining into them, there is also the possibility that changes in the composition of the water may occur during its period of storage in the lake. There are two reasons for this: first, because the residence time of water in lakes is generally longer than in free-flowing rivers, and secondly, because lakes tend to be deeper than rivers so that replacement of dissolved oxygen by recharge from the atmosphere is more difficult. The replacement of dissolved oxygen in lake water is of fundamental importance and it is considered in detail below.

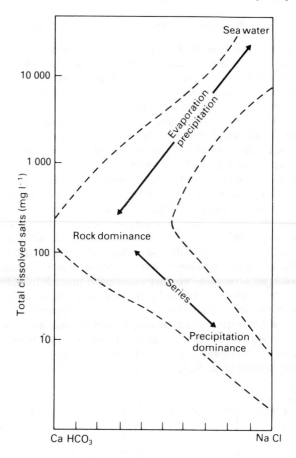

Fig. 3.22 Diagrammatic representation of processes controlling the chemistry of world surface waters (from Gibbs, 1970; copyright 1970 by the American Association for the Advancement of Science).

Table 3.5 The source of Na^+, K^+. Mg^{2+} and Ca^{2+} in various types of rivers (from Gibbs, 1970)

River	River type	Contribution from precipitation (%)	Contribution from rocks (%)
Rio Tefe	Rain dominated	81	19
Ucayali	Rock dominated	4.8	95.2
Rio Grande	Evaporation-crystallisation	0.1	99.9

Dissolved Oxygen, Nutrients and Biogenic Activity in Lakes

The link between the dissolved oxygen concentrations of lake water and biogenic activity can be best understood by considering variations in water temperatures and dissolved oxygen, as a function of season and water depth in a typical temperate lake. Fig. 3.23 shows how the water temperature varies with depth for the four seasons of the year. In winter there is no vertical change in the temperature of the water, which is then said to be isothermal, and in many cases the whole water body is at or near its temperature of maximum density (4°C). With the onset of spring the surface water is warmed by the increased input of solar energy and becomes less dense than the cooler water below, giving the observed two-layer structure. The upper and lower layers are called the epilimnion and hypolimnion respectively, and the transition zone between them, where there is a rapid change in temperature over a small depth interval, is called the thermocline. As surface warming continues through the spring and summer the thermocline moves downwards and the two layer structure becomes more pronounced. In autumn the epilimnion water starts to cool down and at some point becomes cooler, and more. dense, than the water of the hypolimnion. At this time there is substantial vertical mixing or overturning and the water becomes isothermal with depth. During the late autumn-winter period the water continues to cool and mix vertically until it returns to the winter profile (Fig. 3.23).

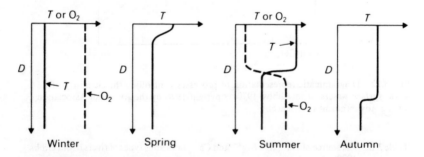

Fig. 3.23 The seasonal variations of water temperature profiles in a temperate lake. The winter and summer profile also show the behaviour of dissolved oxygen in abiological conditions.

The vertical profiles of dissolved oxygen are more difficult to explain since they are produced by not only by physical but also by biological processes. Physical factors alone should be considered, first, for simplicity, and once their role is established other processes can be taken into account. It is assumed that the water at any point in the lake will contain the amount of oxygen expected for equilibrium with the atmosphere at the water temperature. The solubility of oxygen in water, as expressed by Henry's law constant K_H, is known to decrease with increase in temperature, i.e. the dissolution reaction is exothermic. For example, at $10°C$ the value of K_H is 1.65×10^{-3} mol l^{-1} atm^{-1} and at $20°C$ it is 1.34×10^{-3} mol l^{-1} atm^{-1}. Therefore, in the absence of biological or other effects and assuming equilibrium with the atmosphere, dissolved oxygen will be inversely correlated with temperature and the expected vertical O_2 profiles will be as shown in Fig. 3.23. Only the summer and winter profiles are given since those for the other two seasons will lie somewhere in between these extremes. The isothermal water temperature profile in winter leads to a high and vertically homogeneous dissolved oxygen content. With the strong summer thermocline the shape of the dissolved oxygen profile is the reverse of that for temperature.

Having established the effect of water temperature on dissolved oxygen concentrations it is now possible to discuss the all-important role of biological processes. Attention will be focused on the spring-summer profiles since it is then that biological activity is at its most effective. In late autumn and winter plant life is curtailed due to the smaller amount of UV radiation available for photosynthesis and the lower water temperature, and this last factor also inhibits bacterial activity. In spring and summer, plant life in the epilimnion, of which the microscopic phytoplankton are generally the most important, takes up CO_2 from the water and releases oxygen to it as a result of photosynthesis (page 9). Therefore, oxygen is being produced in the upper part of the water column and, although some of this will be lost to the atmosphere across the lake surface and some used in aerobic metabolism (page 140) and respiration, there will be a tendency for supersaturation of dissolved oxygen to occur in the water, i.e. for the near-surface water to contain more O_2 than indicated in Fig. 3.23, where 100% saturation is assumed. Phytoplankton have a very short life cycle of approximately 3 weeks, and on dying they start to sink and decompose by aerobic metabolism, i.e. the photosynthetic reaction is reversed and dissolved oxygen is consumed to break down organic matter to CO_2. The large negative free energy of the decomposition reaction ($\Delta G^\oplus = -2879$ kJ) means that it occurs spontaneously, without the input of energy from an external source (page 21). However, the fact that a reaction is energetically favourable does not mean that it will take place rapidly and, in fact, the breakdown of organic matter in natural waters is kinetically slow without the participation of bacteria. In lakes some decomposition takes place in the epilimnion but it is in the hypolimnion that its effects really become apparent. Near the surface any oxygen used in decomposition can be replaced by the photosynthetic activity of other plankton or by the input of oxygen from the atmosphere. In the hypolimnion neither of these processes is possible and one would expect to find dissolved oxygen concentrations lower than those predicted by the simple physical reasoning illustrated in Fig. 3.23.

The effect of the photosynthesis/decomposition process on lake chemistry will obviously depend on the intensity of such activity, remembering that photosynthesis plays the major role since the formation of cell material is a necessary prerequisite for its decomposition. For surface waters in spring and summer, water temperature and supply of solar radiation are more than adequate for photosynthesis and it is generally agreed that the growth limiting factor is the availability of dissolved plant nutrient substances such as dissolved silica (H_4SiO_4) and the anions PO_4^{3-} (phosphate) and NO_3^- (nitrate). These are required in small amounts by photosynthetic plants and if they are in short supply growth will cease. Therefore, in Equation (1.1) nutrients appear on the left hand side and their constituent atoms, P, N, Si, which are an integral part of the organic material formed, ought to be shown as part of the photosynthetically formed organic molecule. When photosynthesis ceases and Equation (1.1) reverses, i.e. the decomposition reaction, the nutrients are returned to the water and are said to be regenerated.

If nutrient availability is the limiting factor controlling phytoplankton activity, and also the production and consumption of oxygen, all that is required in order to predict the effect of biological activity on the water chemistry is a measure of the nutrient status of the lake. Unfortunately this is very difficult to quantify since it depends not only on the rate of supply of nutrient to the lake from external sources, e.g. inflowing river water, but also the rate at which nutrients are recycled within the lake through regeneration. Because of this lakes are divided into two broad classes: oligotrophic or nutrient poor; and eutrophic or nutrient rich. For oligotrophic lakes the effect of biological activity on, for example, dissolved oxygen profiles will be small relative to that predicted for eutrophic lakes.

Fig. 3.24 compares, in simplified diagrammatic form, the expected vertical profiles for water temperature and dissolved O_2 during the summer season in eutrophic and oligotrophic lakes compared with those with no biological activity as in Fig. 3.23. A comparison of the predictions shown in Fig. 3.24 with the actual summer profiles from several lakes in the English Lake District (Fig. 3.25) indicates that this simplified model satisfactorily explains the observations. For example, Esthwaite resembles Fig. 3.24a and is a typical eutrophic lake, Coniston is similar to the oligotrophic profile in Fig. 3.24b, whereas Wastwater is close to having a dissolved oxygen profile which is the reverse of that for water temperature, as predicted for lakes with little or no biological activity (Fig. 3.24c).

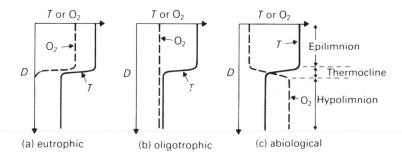

(a) eutrophic   (b) oligotrophic   (c) abiological

Fig. 3.24 Idealised vertical profiles of water temperature and dissolved oxygen in eutrophic, oligotrophic and abiological lakes in summer.

It is apparent that in a eutrophic lake such as Esthwaite, the decomposition of organic matter in the bottom water can reduce the dissolved oxygen to very low levels in the summer. Such waters are called anoxic and are inimical to the survival of animal life, which relies on O_2 for respiration. This situation is only temporary and the seasonal overturn of water in the autumn will bring oxygen-rich surface waters to the bottom of the lake, but in the following spring the photosynthesis/decomposition cycle begins again and from then until late summer the bottom water will again become progressively depleted in dissolved oxygen.

When free dissolved oxygen becomes scarce other electron acceptors are used instead of O_2 in the decomposition of organic matter. For example, nitrate (NO_3^-) can be converted to nitrogen gas, iron (III) to iron (II), and sulphate (SO_4^{2-}) to hydrogen sulphide (H_2S). Some of these reactions are discussed further on page 140, and all are examples of redox reactions, the basic chemistry of which is explained in the Glossary. The photosynthesis/decomposition process (Equation (1.1)) is itself a redox reaction. Many redox reactions are kinetically slow in pure inorganic solutions but in natural systems their reaction rates are substantially enhanced by microbiological activity.

Some redox reductions have important consequences for the chemistry of lake waters. Iron (III) is generally insoluble, except in

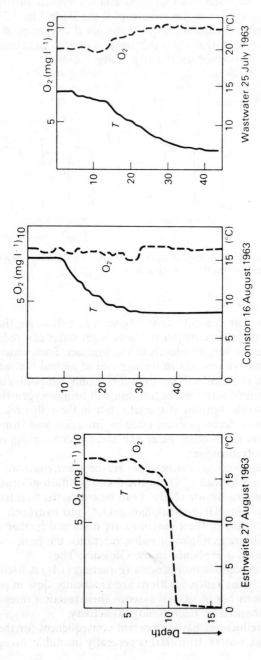

Fig. 3.25 Observed vertical profiles of water temperature and dissolved oxygen in eutrophic, oligotrophic and abiological lakes in summer.

very acid solutions, and therefore exists largely as an oxide or other solid phase. Its reduction to iron (II) causes a great increase in solubility so that iron enters the lake water from the dissolution of solid phases containing Fe^{3+}. The resulting increase in dissolved iron is not important in itself, except that solid phases containing Fe^{3+} invariably contain other substances which are also released to the water along with the iron. For example, iron oxides contain small admixtures of iron phosphate and the release of phosphate to the water may alter the nutrient status of the lake. Another redox reaction which may be important in lakes is the reduction of SO_4^{2-} to H_2S gas (page 140). Hydrogen sulphide not only has an unpleasant pungent odour of rotting vegetables but is also highly toxic to most forms of life.

In many lakes the water is not sufficiently reducing, i.e. it still contains substantial dissolved oxygen, for the above reactions to occur. However, the decomposition of organic matter continues in the bottom sediments, which are not in free contact with oxygen in the water column and in virtually all lakes these sediments become anoxic at some depth. The use of oxidising agents or electron acceptors other than oxygen is discussed further on page 140, with reference to marine sediments, but the general principles are also applicable to freshwater sediments. In some cases the anoxic conditions associated with the formation of reduced species such as Fe^{2+} and H_2S are reached in the water column. However, anoxic conditions, whether in the water column or bottom sediments, are mostly localised because the reduced species are generally rapidly oxidised as soon as oxygen becomes available. The N_2 gas formed by the reduction of NO_3^- is an important exception since the back reaction is kinetically very slow, due to the great strength of the N-N bond.

A very important factor controlling the chemistry of the water in any part of the lake is, therefore, the extent to which decomposition of organic matter has occurred at that point. Since the formation of organic matter is nutrient limited, nutrient supply has been pin-pointed as probably the most important parameter affecting the water chemistry in unpolluted lakes. Lakes in their unspoiled states have only certain fixed amounts of nutrients available for plant growth so that many of them show only moderate O_2 depletion in their bottom waters. However, where there is a large natural supply of nutrients anoxic conditions do develop and alternative electron acceptors to oxygen are utilised. Such a state of affairs can occur quite normally and many lakes show a transition along the sequence oligotrophic → eutrophic → anoxic during the course of their existence.

Anthropogenic Inputs to the Land-based Hydrosphere

Eutrophication in Lakes

Since the factors controlling the chemistry of unpolluted lakes have been established, it is a relatively simple matter to predict the effect of anthropogenic inputs. Important pollutants in the present context are of two types. The first is direct input of plant nutrients, usually as fertiliser-rich run off draining from agricultural land or domestic and industrial detergents. The detergents often contain large concentrations of sodium tripolyphosphate, the function of which is to soften the water (see p. 74) by bonding Ca^{2+} and Mg^{2+} ions as soluble tripolyphosphates. On release to a natural water the tripolyphosphate anion is broken down by hydrolysis to release the phosphate ion, which is an important nutrient in aquatic systems.

The second major source of man-made input to lakes is from sewage effluent which has two effects. The first is that sewage contains large amounts of partially decomposed organic matter, which if allowed to enter a lake or any other water system, adds to the natural pool of decomposing material. The development of anoxic conditions is, therefore, speeded up. Secondly, during the decomposition of the sewage nutrients are regenerated. Therefore, the net effect of these two anthropogenic sources of input is to add to the burden of organic matter which will be decomposed in the lake. This added burden arises either directly from input of sewage effluent or via the *in situ* formation of additional organic matter produced by increased photosynthesis resulting from the augmented nutrient supply. Basically the anthropogenic inputs speed up the progression of a lake along the series oligotrophic → eutrophic → anoxic. In some cases, e.g. Lake Erie, the change to anoxic conditions may be dramatically accelerated.

Although this discussion has concentrated on lakes, the same basic principles apply to other natural waters such as rivers. However, the longer residence times and increased depth of the water in lakes, relative to flowing waters, means that the effects of anthropogenic inputs are often more pronounced in lakes. However, if sufficient nutrient or decomposable organic matter (sewage) is added to a river then similar changes in water chemistry will occur. A good example is the River Thames which for much of the twentieth century has had a long stretch depleted or completely devoid of dissolved oxygen, due to inputs of sewage from the large surrounding urban area. This is shown in Fig. 3.26 which shows the percentage saturation (ratio of measured O_2 to that expected for equilibrium with the atmosphere, expressed as a percentage) as a function of distance up and down stream from London Bridge for various periods during this century. The situation appears to have worsened

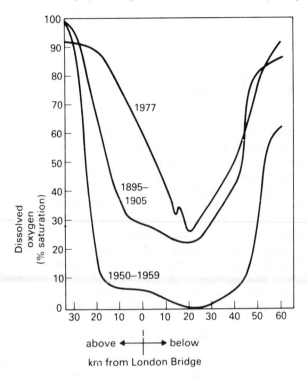

Fig. 3.26 Oxygen saturation levels in the River Thames during the period 1895–1977. The decomposition of sewage by aerobic metabolism (page 140) consumes dissolved oxygen, reducing the concentration below saturation level. Recent higher dissolved oxygen levels reflect reduced sewage inputs.

between the start of the century and the 1950s. However, since then major improvements have been made to the sewage treatment facilities for effluents discharged into the Thames and this has led to a significant increase in dissolved oxygen levels and to an improvement in water quality.

4 The Oceans

In the earth-air-water factory the ocean basins are represented as a huge reaction vessel, which is the ultimate destination of all the dissolved and particulate material derived from the liquid extractor plant. Dissolved materials are transported into the oceans mainly by rivers, groundwaters and atmospheric precipitation and may also be added to a small and unknown extent by marine erosion and submarine volcanicity. The same main sources are responsible for the total particulate supply to the oceans with the contributions from rivers and glaciers predominating to the extent of constituting over 95 % of the known particulate flux. Table 4.1 shows an estimate of the world-wide magnitude of supply processes for both dissolved and particulate material. Rivers are responsible for the bulk of the dissolved and particulate matter being added to the oceans.

Table 4.1 Flux of dissolved and particulate material into the oceans

Source	Dissolved constituents (g yr^{-1})	Particulate matter (g yr^{-1})
Rivers	39×10^{14}	183×10^{14}
Subsurface waters	4.7×10^{14}	4.8×10^{14}
Ice	$<7 \times 10^{14}$	20×10^{14}
Marine erosion		2.5×10^{14}
Atmospheric supply	2.5×10^{14}	6×10^{14}
Volcanicity		1.5×10^{14}

The contributions of particulate matter outweigh dissolved substances by an average ratio of approximately 4:1, but the emphasis will mainly be on the latter in this chapter. The particulate matter supplied by rivers consists mainly of weathered aluminosilicates which, apart from ion-exchange reactions, are effectively inert during their exposure to sea water. Reactions do occur after deposition and burial in the bottom sediments but only after a considerable period of time has elapsed. Possible interactions between particulate material and sea water will be considered later in the chapter, the main purpose of which is to examine the role of sediment-forming chemical processes in controlling the composition of sea water.

The mean compositions of river and sea water were compared in Fig. 2.8 and, it is clear that they differ markedly. Whereas river water is basically a dilute solution of metal cations and bicarbonate ions, sea water is a much more concentrated mixture, in which Na^+ and Cl^- ions predominate. Recognition of this difference in chemistry poses the problem of whether the progressive addition of river water to the ocean basins will eventually alter the composition of sea water, since Fig. 2.8 shows that river water not only contains different proportions of ions but is also more dilute than sea water.

From the hydrological cycle in Fig. 2.6 it is clear that water entering the oceans due to precipitation and run off is balanced by removal due to evaporation. Water used to extract soluble components from rock minerals in the liquid extractor plant is drained into the oceans, from which it is recycled for further weathering reactions after being cleansed by evaporation. In these circumstances it follows that dissolved solids would accumulate in the oceanic reaction vessel, since negligible quantities of dissolved solids would be lost during bubble bursting compared to the amounts added by water percolating through the liquid extractor plant. Confirmation of this is achieved by comparing the rates of river supply of various dissolved components with their rates of removal during the bubble-bursting process (Table 4.2). Rates of addition greatly exceed rates of removal, except for Na^+ and Cl^-, which implies that dissolved constituents accumulate in the oceanic reaction vessel and sea water is becoming more saline.

Table 4.2 Comparison of atmospherically-cycled salts and annual river load (from Garrels and Mackenzie, 1971)

Constituent	Annual flux (g)	
	Salts cycled atmospherically by bubble bursting	Dissolved river load
Na^+	7.33×10^{13}	20.7×10^{13}
K^+	1.11×10^{13}	7.4×10^{13}
Mg^{2+}	0.99×10^{13}	13.3×10^{13}
Ca^{2+}	0.33×10^{13}	48.8×10^{13}
Cl^-	14.0×10^{13}	25.4×10^{13}
SO_4^{2-}	2.15×10^{13}	36.7×10^{13}
HCO_3^-	0.44×10^{13}	190.2×10^{13}

Rates of weathering processes are mostly slow as measured on a human time-scale so that it is relevant to find out on what scale measurable natural changes in oceanic chemistry could have occurred. This problem can be answered by considering the residence times for each of the major dissolved species in sea water (Table 4.3). These values indicate how long it would take for the river flux of any species to equal the present oceanic content of that species. As one might expect the residence times of dissolved species are so long that there is virtually no chance of analytical information being used to detect changes in sea water chemistry over the period of time, probably only the last century, for which reliable information is available.

Table 4.3   Residence times of major sea-water constituents (from Garrels and Mackenzie,
1971)

Constituent	Residence time (years)
Na^+	7.0×10^7
K^+	7.0×10^6
Mg^{2+}	1.4×10^7
Ca^{2+}	1.2×10^6
HCO_3^-	1.0×10^5
SO_4^{2-}	1.0×10^7
Cl^-	1.0×10^8
SiO_2	2.0×10^4

It is necessary, therefore, to look at the geological record to provide evidence for the changes in, or constancy of, sea water composition. Definite evidence is difficult to find but the little there is, together with much circumstantial evidence, does suggest that the chemical composition of sea water has remained appreciably constant for the last $10^8 - 10^9$ years. For example, some families and even species of organisms have existed for approximately the last 5 $\times 10^8$ years, which is an unlikely occurrence if sea water had undergone substantial modification. Over this same time span the skeletal composition of fossil brachiopods, a type of bivalve, is chemically similar to modern mollusc shells, in respect of the ratios of the oxygen isotopes ^{18}O and ^{16}O and the contents of the trace elements Sr and Mg in the $CaCO_3$. There is some negative evidence too, which essentially places limits on the permitted range of sea-water variation. For silica concentrations in excess of 25 mg l^{-1} SiO_2, a sea water at pH 8 would precipitate the magnesium silicate sepiolite $Mg_2Si_3O_6(OH)_4$. In fact this mineral is rare in either ancient or modern marine sedimentary rocks and it is, therefore, concluded that the silica concentration of the oceans either did not reach this level. or did not persist there for any length of time. A similar argument has been applied to the mineral Mg(OH)$_2$ (brucite) whose precipitation from sea water would occur if the pH was maintained at 10. However, brucite never occurs extensively in marine sediments which is an indication that the pH of sea water has been confined to a smaller range of variation. Another piece of evidence comes from a comparison of the boron contents of shales as a function of their geological age (Fig. 4.1). The amount of boron in these clay-rich rocks is related to the concentration in sea water at the time of sediment deposition. The lack of variation with time indicates that the boron concentration of sea water has been relatively constant over the last $2.5 - 3.0 \times 10^9$ years.

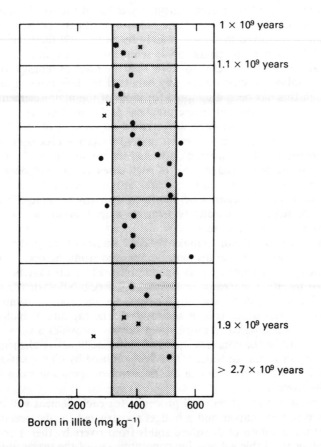

Fig. 4.1 The boron content of ancient rocks (> 10^9 yrs old) shows little variation with age and is similar to the range found in younger rocks (shown by the shaded band).

These isolated observations are hardly conclusive but are given substantial support by the general similarity of ancient and modern marine sediments, both in terms of the mineralogy of particular sediment types and their relative abundance. The implication is that the chemistry of, and chemical processes in, the oceans have been broadly similar over a substantial period of geological history. The importance of this conclusion is that the oceans cannot be a mere accumulator for the dissolved products of continental weathering, but are a dynamic system in which a variety of chemical and physical processes must operate to remove dissolved materials as fast as they are added. Only in this way can the ocean maintain a constant composition.

This dynamic view of ocean chemistry is called the steady state model and its prerequisite is that a balance must exist between the addition of dissolved materials, mainly by rivers, and their removal into solid phases by precipitation as sediments. An elegant justification of this model is to show that the rates of removal of the major dissolved species are roughly balanced by their rates of input. Although this has been attempted for most of the major elements in sea water, it does present severe practical problems. Surface processes have generally been well-studied on account of their accessibility but this is not true for many submarine processes. This point emerges with considerable force in the attempt to make a silica budget for the ocean (page 118). In such cases as this, not only are accurate estimates of the rates of removal impossible but the processes themselves cannot be identified with any certainty. Overall, it is fair to regard the oceans as being no more than in an approximately steady state.

Although it is difficult to make detailed studies of the pathways by which dissolved materials may be removed from the ocean, two different types of pathway may be identified. Elements may be removed by either inorganic processes, e.g. precipitation or reaction with solid phases, or biological processes, e.g. incorporation into skeletal material which is later sedimented. The rapidity of biological processes, as compared to inorganic processes, provides a very effective limit to the concentrations of biologically active elements which may occur in sea water. Thus SiO_2, utilised by phytoplankton, has the lowest residence time of all the dissolved species in sea water given in Table 4.3.

The most important removal processes for each element will be discussed in this chapter and a budget constructed for the ocean, to balance the addition of dissolved solids from rivers by their removal into sediments. Table 4.4 lists the concentrations of the major ions in river water and, given that the annual world-wide volume of river discharge is 3.3×10^{16} litres, it is a simple matter to calculate the

Table 4.4 Annual discharge of dissolved species by rivers (from Garrels and Mackenzie, 1971)

Constituent	Concentration in river water ($m\ mol\ l^{-1}$)	Annual discharge ($10^{10}\ mol\ yr^{-1}$)
Cl^-	0.220	715
Na^+	0.274	900
Mg^{2+}	0.171	554
SO_4^{2-}	0.117	382
K^+	0.059	189
Ca^{2+}	0.375	1220
HCO_3^-	0.958	3118
SiO_2	0.218	710

total number of moles of each element added to the oceans over one year (see Table 4.4). In order to maintain a steady state ocean, removal processes must be capable of extracting these excess dissolved constituents. The necessary chemical mass balance reactions or budgets are derived by removing the appropriate quantities of each element that are needed to form common minerals in marine sediments. The general approach is adapted from that of Mackenzie and Garrels (1966).

Evaporite Formation

All the incoming chloride and sulphate ions can be removed from the oceanic reaction vessel by precipitation as the minerals halite (Na Cl) and gypsum (Ca $SO_4.2H_2O$). In terms of the chemical budgets the following reactions are predicted:

$$Na^+ + Cl^- \rightarrow NaCl$$
$$Ca^{2+} + SO_4^{2-} + 2H_2O \rightarrow CaSO_4.2H_2O$$

During the precipitation of both minerals equimolar amounts of cations and anions are abstracted into the solid phases. Therefore, if all the incoming chloride is precipitated as halite and all the sulphate as gypsum, there will be a surplus of $(900 - 715) \times 10^{10}$ moles of Na^+ and $(1220 - 382) \times 10^{10}$ moles of Ca^{2+} (Table 4.5). Other processes operate to remove these surpluses by the formation of additional Na^+ and Ca^{2+} − bearing phases.

To investigate the conditions in which halite and gypsum form naturally, the residence times of the four ions involved should be noted, and it is found that they are all comparatively large (Table 4.3). Their high sea-water concentrations indicate the difficulty with which such elements form solid phases under surface conditions.

Table 4.5 Budget for the removal of river-borne dissolved constituents from the oceans

	Balance of constituents (10^{10} mol)							
	SO_4^{2-}	Cl^-	Ca^{2+}	Mg^{2+}	Na^+	K^+	SiO_2	HCO_3^-
Annual addition by rivers	382	715	1220	554	900	189	710	3118
Removal processes:								
Evaporative processes			838	554	185	189	710	3118
Carbonate formation				510	185	189	710	1354
Silica formation				510	185	189	530	1354
Interactions between clays and sea water				500			530	958
Reverse weathering							430	−42

Evaporative processes
$715Na^+ + 715Cl^- \rightarrow 715NaCl$
$382Ca^{2+} + 382SO_4^{2-} \rightarrow 382CaSO_4$

Carbonate formation
$838Ca^{2+} + 1676HCO_3^- \rightarrow 838CaCO_3 + 838CO_2 + 838H_2O$
$44Mg^{2+} + 88HCO_3^- \rightarrow 44MgCO_3 + 44CO_2 + 44H_2O$

Silica formation
$180H_4SiO_4 \rightarrow 180SiO_2 + 180H_2O$

Interactions between clays and sea water
$Ca\text{-clay} + 189K^+ \rightarrow 95K_2\text{-clay} + 95Ca^{2+}$
$Ca\text{-clay} + 185Na^+ \rightarrow 93Na_2\text{-clay} + 93Ca^{2+}$
$188Ca^{2+} + 376HCO_3^- \rightarrow 188CaCO_3 + 188CO_2 + 188H_2O$
$10Mg^{2+} + 20HCO_3^- \rightarrow 10MgCO_3 + 10CO_2 + 10H_2O$

Reverse weathering
$100Al_2Si_2O_5(OH)_4 + 500Mg^{2+} + 100SiO_2 + 1000HCO_3^-$
$\rightarrow 100Mg_5Al_2Si_3O_{10}(OH)_8 + 1000CO_2 + 300H_2O$

Indeed the precipitation of NaCl and $CaSO_4.2H_2O$ is essentially independent of pH and oxidation/reduction, so that the only way these minerals can be formed is by inducing supersaturation through evaporation. The problems with evaporative precipitation are that gypsum and halite are both readily soluble in water at surface temperatures and pressures (approximately 2.4 and 357 g l^{-1} respectively) and so both minerals are greatly undersaturated in sea water. This can be illustrated by a simple laboratory experiment which gives a valuable insight into the formation of evaporite minerals.

Usiglio's experiment

In 1849 the Italian chemist Usiglio evaporated a sample of water from the Mediterranean to dryness, collecting, identifying and weighing the products. One litre of evaporated sea water gave the products shown in Table 4.6. First to precipitate is $CaCO_3$, followed by $CaSO_4.2H_2O$ and then by NaCl, whose formation is overlapped by deposition of $MgSO_4$ and $MgCl_2$. The final stages of the evaporation give variable results, and usually consist of simple chlorides and bromides together with more complex salts. Even ignoring those salts which precipitate from the final brine, known as bittern salts, the experiment gives much useful data.

Firstly, large volumes of sea water (approximately 47%) must be, evaporated even before the first mineral, $CaCO_3$, appears. Such conditions cannot develop in the open ocean, where the overall rate of water loss by evaporation is balanced by continental run off (page 29) and where effective mixing in the surface layer is maintained by winds, waves and currents. However, the effect of mixing may be reduced or even eliminated in any partially or wholly-isolated portion of sea water. Evaporation may then predominate locally over the resupply of water by precipitation, surface run off and sea-water influx. On a small scale such conditions are common, e.g. salt-fringed pools occur above the high-tide level on many rocky beaches, but the exclusion of resupply processes becomes increasingly difficult on a larger scale. This in turn means that higher evaporation rates are needed to ensure the predominance of evaporation over resupply, which effectively restricts large scale evaporite formation to a narrow latitude band on either side of the equator where the climate is favourable.

Secondly, the laboratory experiment predicts the sequence and relative thickness of the different evaporite minerals. The evaporation of an enclosed ocean basin would successively precipitate $CaCO_3$, $CaSO_4.2H_2O$ and NaCl in the proportions 1:15:254 by weight. However, the amounts of $CaCO_3$ formed this way are relatively

Table 4.6 Usiglio's experiment with the precipitation of salts from sea water at 40°C.

Volume of sea water (litres)	Density (g cm^{-3})	Salinity• (‰)	Amount of salt (g)						
			CaCO$_3$	CaSO$_4$·H$_2$O	NaCl	MgSO$_4$	MgCl$_2$	NaBr	KCl
1.00	1.03	37.5							
0.53	1.05	68.5	0.0642						
0.19	1.13	157.1	0.053	0.5600					
0.112	1.20	274.0		0.906					
0.095	1.21	291.0		0.0508	3.2614	0.004	0.008		
0.039	1.24	325.7		0.2176	17.546	0.0392	0.079	0.073	
0.0302	1.26	339.5		0.0144	2.624	0.0174	0.015	0.358	
0.0162	1.31	399.0			3.676	0.5636	0.0514	0.1138	
Total precipitated			0.1172	1.7488	27.1074	0.6242	0.1532	0.2224	
Salts in bittern					2.5885	1.8545	3.1640	0.3300	0.534
Total solids			0.1172	1.7488	29.696	2.479	3.317	0.552	0.534

• Expressed as parts per thousands(‰)

small and are mostly insignificant compared to $CaCO_3$ deposition as skeletal materials (page 112). In these circumstances it is usual to consider $CaSO_4.2H_2O$ as the first true evaporite mineral, and it is not surprising to find that this is the most frequently developed mineral in ancient evaporite sediments. Not only is it the first to be formed, but it is also the least soluble and so the most likely to survive subsequent weathering processes. One would also expect that if evaporation continues for a sufficient length of time, a thick sequence of halite may develop. Some ancient evaporite sediments certainly do show such thickness of halite. Finally, the last fraction of sea water is the most difficult to evaporate, and, therefore, it is anticipated that minerals such as sylvite (KCl) will be relatively rare in ancient evaporites. Again, this observation is supported by the field study of evaporite sediments.

The Barred Basin Model

Despite general agreement between the predictions from laboratory evaporation of sea water and the observations of ancient evaporites, considerable differences exist. The two main disparities are firstly that some ancient evaporite deposits are over 1 km thick, which implies the evaporation of a layer of sea water over 100 km thick (the maximum depth of the oceans being approximately 11 km) and secondly that the ratio of gypsum to halite is very much higher than that predicted. These problems can be resolved by evaporating a bay or arm of the sea, which has a restricted connection with the open ocean rather than being totally enclosed. The typical model (Fig. 4.2) is referred to as a barred basin. The surface arrows indicate the paths of inflowing water from the ocean, which is evaporated and becomes denser towards the distant end of the basin. Here the dense brines sink and flow back along the bottom of the basin. At first the dense

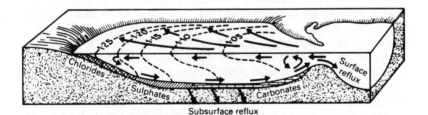

Fig. 4.2 Model evaporite basin. Numbered lines represent contours of equal water densities that increase away from the open ocean. Surface arrows show paths of influxing water. At depth, there is a reverse flow of dense brine along the sea floor. Reflux takes place by seepage through the sediments and over the bar.

brines displace fresh sea water from the basin, but the water
displaced gradually becomes denser itself. Provided denser brines can
be generated at the far end of the basin than are present in its lower
regions, the latter can always be pushed over the sill and out of the
basin. This process is known as reflux, and it provides an effective
mechanism for introducing fresh sea water into the basin to replace
that lost by evaporation. Therefore, thick sequences of evaporites do
not imply the evaporation of unreasonably large initial depths of
water.

The barred basin model also supplies a solution for the bias in
ancient evaporites towards gypsum, rather than halite. Fig. 4.2
suggests that the net loss of water from the basin represents the
difference between evaporation and influx. Given the appropriate
balance between evaporation and influx, it would be possible to keep
a brine precipitating gypsum for long periods without it over
becoming sufficiently concentrated to precipitate halite. Clearly the
ratios of different evaporite minerals reflect kinetic factors, i.e.
primarily rate of evaporation and rate of influx of sea water, which
may vary irregularly. For example, some evaporite deposits show
sequences such as $CaCO_3 - CaSO_4 . 2H_2O - NaCl - CaCO_3$
$- CaSO_4 . 2H_2O$. This indicates an early dominance of evaporation
over influx, as the brines increase in density sufficiently to form
NaCl, then a reduction in brine density due to a sudden influx of
water which returns the basin back to the first stage of $CaCO_3$
deposition. Each separate sequence of minerals representing a
gradual increase in brine density is referred to as a cycle. Evaporite
sediments often show many such cycles, reflecting variations in the
relative magnitudes of the kinetic controls.

There is one further complication. Where a density gradient exists
in the basin, the precipitation of different evaporite minerals may
occur simultaneously in different parts of the basin. The situation
shown in Fig. 4.2 is found in most ancient evaporites at some point
in their evolution, depending on whether the density gradient in the
basin overlaps the range required for the deposition of two or more
minerals. This spatial variation of evaporites must be linked to the
temporal variations which give rise to cycles, to explain the
complexities of evaporite sedimentation. Fig. 4.3 shows the history of
an evaporite basin which experiences one major salinity reversal, due
to a sudden influx of sea water, perhaps because of high tides or
temporary collapse of the sill. The water salinity and sediment
mineralogy are shown varying both temporally and spatially. During
the intial evaporation phase, a salinity gradient is established which
enables NaCl to be formed at the distant end of the basin. The
subsequent inflow of sea water reduces the salinity gradient and
changes the mineralogy of the evaporites forming in some parts of

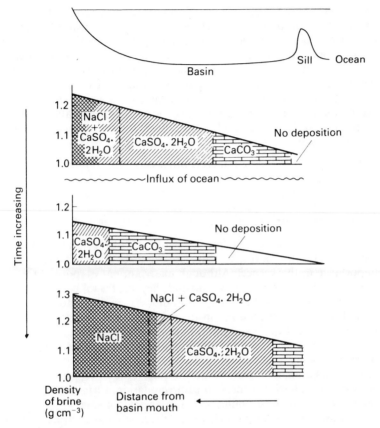

Fig. 4.3 Diagrammatic representation of the temporal and spatial variations in an evaporite basin which has a sudden influx of fresh sea water.

the basin. This event marks the end of one cycle and the beginning of the next, in which evaporation rates are rather more dominant since gypsum forms even at the mouth of the basin. It should be noted that each locality in the basin exhibits its own evaporite cycle, reflecting the salinity history at that point, but that sediments forming at the same time may be mineralogically different.

Evaporites as Removal Mechanisms for Na^+, Cl^-, Ca^{2+} and SO_4^{2-}

The intricate mechanisms of evaporite formation are almost exclusively deduced from a study of ancient evaporite sediments. It is a striking fact that there are no modern areas of evaporite formation analogous to the barred basins inferred from the study of ancient evaporites. There are some areas of evaporite deposition but these occur mainly in the Persian Gulf, where the coast is fringed by an extensive area of carbonate sediments lying only 1 m above high tide. These supratidal flats are called sabkhas and may be approximately 25 km in width. Periodic tidal incursions introduce sea water on to the flats and into the pore spaces in the sediment: Evaporation of the sea water contained in the sediment forms brines which usually precipitate calcium sulphate minerals and occasionally halite. The evaporites produced on sabkhas differ from basinal evaporites in a number of ways:

(1) The calcium sulphate minerals often occur as nodules in a carbonate matrix
(2) The sediments show evidence of sub-aerial exposure, such as desiccation cracks and wind-blown dust
(3) The tidal incursions may cause erosion features.

Clearly the sabkhas are modern areas of Ca^{2+}, SO_4^{2-}, Na^+ and Cl^- removal from the oceans, but they appear to be of only minor importance in this respect. Although many ancient evaporites are now being re-examined to see whether they show sabkha or basinal features, it does seem that there is no modern analogue for the barred basins which occurred in the geological past. Some possible reasons for this can be suggested by considering the conditions necessary to develop barred basin evaporites, i.e. the presence of a basin possessing a restricted connection with the open sea, which is located in a climatic region where the rate of water loss by evaporation can exceed the sum of precipitation, surface run off into the basin and sea-water influx. Even in an arid climate with minimal run off and rainfall, evaporation is rarely capable of abstracting more than 2 m yr^{-1} of water. This imposes a very tight constraint upon sea-water influx; in terms of basin geometry it means a landlocked basin connected to the open ocean by a narrow channel. Basins of this sort currently contain the Red Sea and the Mediterranean, yet in neither case are concentrated brines developing. The Red Sea basin (Fig. 4.4) lies in the more favourable climatic regime, having annual evaporation rate 1.3 m yr^{-1} as compared to 1.0 m yr^{-1} in the Mediterranean, and even has a partially blocked channel, whose depth is approximately 10 m as compared to basin depths of 2000 m. Despite this, sea-water influx is effective enough as a diluent to prevent gypsum and halite precipitation. It would seem that extreme restrictions on influx are necessary, by having an even narrower channel or a channel blockage which extended to within a few meters of mean sea-water level. Geological evidence on the frequency of ancient evaporite indicates that such conditions have never been common. Instead evaporite deposition has tended to occur at

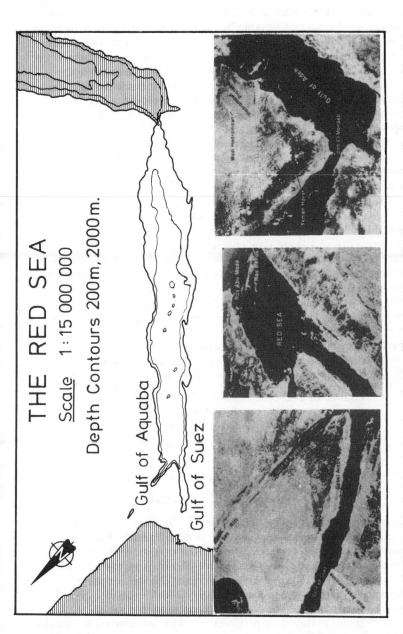

Fig. 4.4 Outline map of the Red Sea showing its enclosed and basinal nature. Satellite photographs from Gemini XI show the distant end of the basin terminating in the Gulf of Suez and Gulf of Aquaba, a view down the length of the Red Sea, and the connection with the Gulf of Aden (Indian Ocean) through the straits at Bab el Mandab (photographs S-66-54893, S-66-63479, S-66-54536 courtesy of NASA).

sporadic intervals throughout earth history. The reasons for this are not fully comprehended although they are in part related to there being a probability of suitable basins forming during the separation (rifting) of continents. During the early stages of separation, tension develops and is accompanied by crustal thinning. As the floor of the rift axis sinks to sea-level the conditions become favourable for extensive intermittent marine incursions. These linear basins along the margins of the separating continents are potential sites of evaporite development. However, as the continents drift further apart, the basins become larger and it becomes less likely that a sufficiently constrained connection with the ocean would exist. The maintenance of such constrained conditions is only possible for geologically brief periods of time, before separation is complete and a new ocean is formed.

Evaporites associated with the preliminary rifting of Africa and America, approximately 160–220 million years ago, formed along the margins of the two continents as the Atlantic Ocean opened. These evaporites (Fig. 4.5) are now covered by coastal marine sediments derived from continental weathering. The axial length of the newly-opened Atlantic Ocean was approximately 4000 km and evaporites probably exist along a substantial proportion of the continental margins to this ocean. Evaporites formed in this way are usually very large. Kinsman (1975) has estimated that these continental margin evaporites may have widths of 50–300 km and thickness of 1–7 km. Each evaporite body has been separated into two halves by the occurrence of rifting. Assuming a typical width of 200 km and thickness of 4 km, the total volume of evaporites can be estimated.

$$\begin{aligned} \text{Volume of evaporite} &= 4000 \times 200 \times 4 \times 2 \\ &= 6.4 \times 10^6 \text{ km}^3 \\ &= 6.4 \times 10^{21} \text{ cm}^3 \end{aligned}$$

If all this volume is NaCl of density 2.2 g cm^{-3}

$$\begin{aligned} \text{Weight of NaCl} &= 1.4 \times 10^{22} \text{ g} \\ &= 2.4 \times 10^{20} \text{ moles NaCl.} \end{aligned}$$

This represents approximately a third of the total NaCl in the oceans. However, the evaporites are believed to have been precipitated over a period of 10–20 million years. Over the same time interval as the evaporites formed, the river supply of Na^+ and Cl^- would be at the rate of 9.0×10^{12} and $7.2 \times 10^{12} \text{ mol yr}^{-1}$ respectively. In 10^7 years approximately 7×10^{19} moles of NaCl would be supplied to the ocean to compensate for the 2.4 $\times 10^{20}$ moles removed by precipitation. The net removal is, therefore, approximately 1.7×10^{20} moles, which represents about 25 % of the

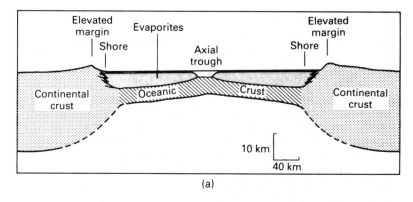

(a)

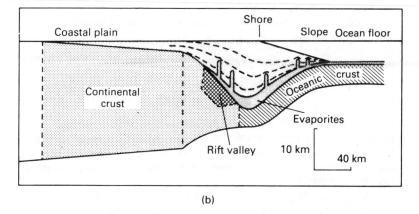

(b)

Fig. 4.5 The formation of evaporites in the ocean between rifting continents (i) and the subsequent burial history as layers of sediment (dotted lines) are added (ii). Weight of added sediment forces salt-domes into overlying sediment (from Kinsman, 1975).

present NaCl content of the oceans. This constitutes a very rapid removal of the ions involved in evaporite formation and suggests that large salinity variations may have accompanied phases of widespread evaporite formation. Clearly the view of the steady state ocean may have to be modified in some respects. Although smaller scale processes may maintain an approximate steady state it is probable that temporary deviations from the steady state value have intermittently occurred. However, on a long-term basis, the average annual rate of evaporite precipitation is capable of removing a large proportion of the river discharge of Ca^{2+} and Na^+ together with all the Cl^- and SO_4^{2-} (Table 4.5).

Carbonate Formation

The budget calculation for the oceanic reactor vessel is taken to its second stage by considering the effect of precipitating carbonates:

$$Ca^{2+} + 2HCO_3^- \rightarrow CaCO_3 + CO_2 + H_2O$$

In this simple approach all the Ca^{2+} (838×10^{10} moles) remaining after evaporite formation is precipitated as $CaCO_3$. Therefore, 838×10^{10} moles of Ca^{2+} react with 1676×10^{10} moles HCO_3^- to produce 838×10^{10} moles of $CaCO_3$. However, studies of modern carbonate sediments indicate that they contain significant proportions of magnesium in isomorphous substitution for calcium in the $CaCO_3$ crystal lattice. The composition of the $CaCO_3$ consists of approximately 95% Ca^{2+} and 5% Mg^{2+} on average and, therefore, an amount of Mg^{2+} equal to 5% of the Ca^{2+} is also removed from the budget as magnesium carbonate (Table 4.5):

$$Mg^{2+} + 2HCO_3^- \rightarrow MgCO_3 + CO_2 + H_2O$$

or in moles:

$$(44 \times 10^{10})Mg^{2+} (88 \times 10^{10})HCO_3^- \rightarrow (44 \times 10^{10})MgCO_3$$
$$+ (44 \times 10^{10})CO_2 + (44 \times 10^{10})H_2O$$

Therefore, carbonate deposition returns carbon dioxide to the atmosphere. The bicarbonate ions were originally delivered by the reactions of silicate minerals and $CaCO_3$ with atmospheric CO_2 and are returned as CO_2 to the atmosphere through the precipitation of calcium carbonate.

Under what sort of conditions can sea water be made to precipitate carbonate minerals? A careful study of modern oceanic sediments indicates that undisputed cases of the inorganic precipitation of $CaCO_3$ under normal marine conditions are rare. However, calcium carbonate shells survive on beaches. Presumably, if sea water were greatly undersaturated with respect to $CaCO_3$, these shells would dissolve. Calcium carbonate saturation in sea water can be crudely explained by a simple thermodynamic model.

Calcium Carbonate Saturation in Sea water

Analysis of average surface sea water shows that the calcium concentration is 0.01 mol l⁻¹ and the carbonate ion concentration is 0.00027 mol l⁻¹. The ion product $[Ca^{2+}][CO_3^{2-}]$ is, therefore, 2.7×10^{-6}, yet the solubility product of calcite is 4.5×10^{-9}. This implies that sea water is greatly supersaturated with respect to $CaCO_3$ as its ion product is more than 10^3 times as large. However, this calculation is only valid for dilute solutions and unfortunately sea water is thermodynamically a concentrated and complex solution. The differences between dilute and concentrated solutions occur in two main ways. Firstly, ions within a concentrated solution must necessarily be closer together and so various forms of electrostatic interaction can occur. The effect of this is to effectively reduce the numbers of ions which are at any time available for reaction and in order to predict the solution behaviour correctly, calculations must account for these inactive ions. Therefore, an active concentration or activity which is related to true concentration is defined as follows:

Activity = Concentration $\times \gamma$

where γ is the activity coefficient, with values ranging from 0 to 1, and is a measure of the proportion of active ions. Values of γ can be determined by experiment or can be calculated from inorganic solution theory. Moderate agreement is obtained from both methods and activity coefficients generally show a steady decrease with increasingly concentrated solutions. As might be expected, the more concentrated a solution, the closer ions are forced together with correspondingly increased opportunities for electrostatic interaction. γ for calcium is approximately 0.26 and for carbonate ions 0.20; both values being for ions in a solution of similar ionic strength to sea water. Therefore, the ion product, which is written in terms of concentration, can be replaced by the ion activity product.

$$
\begin{aligned}
\text{Ion activity product} &= A_{Ca} \times A_{CO_3^{2-}} \\
&= [Ca^{2+}]\gamma_{Ca^{2+}} \times [CO_3^{2-}]\gamma_{CO_3^{2-}} \\
&= 10 \times 10^{-3} \times 0.26 \times 0.27 \times 10^{-3} \times 0.20 \\
&= 1.5 \times 10^{-7}.
\end{aligned}
$$

This value is still almost 30 times larger than the equilibrium solubility product so clearly there are either additional corrections to be made to the model or else sea water is greatly supersaturated with respect to $CaCO_3$.

The second difference between dilute and concentrated solutions is that corrections must be made in the latter to account for associations between ions. This is quite distinct from the rather vague type of interaction corrected for previously. Association creates discrete entities involving fixed proportions of the ions concerned. For example, Ca^{2+} forms an association, called an ion pair, with the abundant SO_4^{2-} ion in sea water. This species is written $CaSO_4^0$ and has no net charge. Other examples of important ion pairs in sea water are $CaHCO_3^+$ and KSO_4^-. It must be emphasised that these species are not solids but are truly present in solution. Table 4.7 lists the dissociation constants for the ion pairs between all the major cations and anions in sea water.

Table 4.7 Ion pair dissociation constants for the major ions in natural waters at 25°C and 1 atmosphere (from Berner, 1971)

Cations	Anions			
	HCO_3^-	CO_3^{2-}	SO_4^{2-}	Cl^-
K^+	> 1	> 1	$10^{-0.96}$	$\gg 1$
Na^+	$10^{0.25}$	$10^{-1.3}$	$10^{-1.1}$	$\gg 1$
Mg^{2+}	$10^{-1.16}$	$10^{-3.4}$	$10^{-2.4}$	$\gg 1$
Ca^{2+}	$10^{-1.26}$	$10^{-3.2}$	$10^{-2.3}$	$\gg 1$

$$K_D = \frac{A_{\text{metal ion}} \times A_{\text{anion}}}{A_{\text{ion pair}}}$$

Clearly the larger the value of K_D, the higher the proportion of ions present as isolated species and the less important the ion pair. For Ca^{2+} ions the important ion pairs occur with the CO_3^{2-} and SO_4^{2-} ions and for CO_3^{2-} ions important ion pairs are formed with Ca^{2+}, Mg^{2+} and Na^+. Calculation of the concentrations of the different sulphate ion pairs, for example, involves the solution of many simultaneous equations since one has to consider not only the equilibrium constants but also the concentrations of all the other ions with which pairing can occur. Therefore, although the $MgSO_4^0$ ion pair forms more readily than the $NaSO_4^-$ pair, the larger concentration of Na^+ ions, as compared to Mg^{2+}, causes both cations to pair with a similar number of sulphate ions. Solution of these simultaneous equations would show that 91 % of the total dissolved calcium is present as the free ion, whilst 8 % is present as the $CaSO_4^0$ ion pair and 1 % as $CaHCO_3^+$. For carbonate ions, 10 % are present as the free ion with 64 % as $MgCO_3^0$, 7 % as $CaCO_3^0$ and 19 % as $NaCO_3^-$.

In the ion product calculation, it is apparent that significant proportions of the analytically determined calcium and carbonate ions are present in ion pairs. Only the concentrations of free calcium and free carbonate must be considered and, therefore, the final corrected ion product has the value:

$$A_{Ca^{2+}} \times A_{CO_3^{2-}} = 1.5 \times 10^{-7} \times 0.91 \times 0.10$$
$$= 1.4 \times 10^{-8}$$

When compared to the solubility product of $CaCO_3$ the above calculations predict that sea water is approximately 3 times supersaturated with respect to $CaCO_3$. However, it is equally clear that sea water is a complex solution and that our thermodynamic model involves a number of possible unjustified assumptions. Overall it is fair to conclude that surface sea water is saturated with $CaCO_3$

and therefore, to anticipate that slight deviations in conditions may allow $CaCO_3$ precipitation at one place and $CaCO_3$ dissolution, e.g. in deep waters, in another. The different environments in which deposition and dissolution occur are discussed in the following sections.

Controls of Calcium Carbonate Formation

The present-day carbonate output from the oceans into marine sediments consists almost entirely of two different crystal forms, or polymorphs[†], of $CaCO_3$, calcite and aragonite. Since surface sea water is oversaturated with respect to $CaCO_3$. It might be expected that widespread $CaCO_3$ precipitation would occur in areas of shallow water, such as the continental shelves. In fact, examples of strictly inorganic precipitation are limited and organisms are responsible for nearly all $CaCO_3$ precipitation in the oceans today.

One of the most widely accepted examples of inorganic precipitation from sea water is the aragonite ooids. Ooids consist of crystals of aragonite surrounding a central nucleus, which is usually a $CaCO_3$ skeletal fragment. Concentric layers of aragonite build up around the nucleus to form a spherical body, usually 0.2–2.0 mm in diameter. Ooids occur in subtropical and tropical areas of $CaCO_3$ deposition where strong wave or current action produces turbulent water conditions locally. High energy environments of this type are important because ooid formation is dependent on temporary suspension of the grains in sea water. Favourable conditions for $CaCO_3$ deposition are created by slight evaporation of the shallow water, which increases the concentrations of Ca^{2+} and CO_3^{2-} ions, hence increasing the ion activity product and the degree of oversaturation of calcium carbonate. In contrast to the overlying sea water, the sea water trapped in the buried sediments cannot be evaporated and does not reach a sufficient degree of supersaturation to induce precipitation. Growth of the ooids, therefore, only occurs during suspension and not whilst the grains are present in the bottom sediments. In fact, ooids in the sediments are subject to abrasion, because turbulence keeps the grains in continual motion, jostling and rubbing against their neighbours. There is, therefore, a continual conflict between abrasion and growth, as the ooid is transported from the sediment to the overlying sea water. The balance of this conflict depends upon ooid size. Small ooids can easily be suspended by weak turbulence and so grow rapidly. Their increasing size, however, makes suspension and growth more difficult. The narrow size distribution of ooids is an expression of the balance between abrasion and growth which occurs when suspension and sedimentation occur for similar lengths of time.

The conflict between growth and abrasion is only terminated when the grain becomes too deeply buried for turbulence to cause either abrasion or suspension.

However, ooid development and other examples of inorganic precipitation are of minor significance in maintaining the $CaCO_3$ balance of the ocean. Most $CaCO_3$ production occurs with the formation of skeletal material by photosynthetic organisms. The environments in which prolific organic productivity occurs are usually characterised by warm, clear, shallow water (Fig. 4.6). Photosynthesis and, therefore, productivity are more rapid in warm waters, whilst solar radiation does not effectively penetrate to depths of greater than 20 m in clear water, or less in turbid waters. These conditions are satisfied to some extent in most shallow surface waters and $CaCO_3$ skeletal material, sedimented after the organism's death, is a common component of most marine sediments. However, adjacent to the continents the input of weathered debris by rivers can inhibit productivity by reducing the depth of light penetration in sea water. The most prolific environments of $CaCO_3$ production

Fig. 4.6 A modern carbonate forming environment, Bimini Lagoon, Bahamas. The sea grass *Thalassia* is shown together with the carbonate secreting organisms *Porites* (a coral) and *Penicillus* (algae). The sediment consists of broken down skeletal material (courtesy of R. G. C. Bathurst from *Carbonate Sediments and their Diagenesis*, Elsevier).

occur in tropical or subtropical waters and adjacent to flat or arid hinterlands, where river input is minimal, or are located on an isolated platform rising from the oceanic depths.

The Grand Bahama Bank (Fig. 4.7) is a modern example of a carbonate platform and is characterised by a variety of different carbonate sediments. Ooids form at the platform margin where tidal currents are focused and where the productivity of calcareous organisms is high due to the rise of nutrient-rich waters from the ocean depths. Further on the platform turbulence is lower and finer sediments can settle from the water, such as the faecal pellets of carbonate mud ingesting organisms and broken fragments of skeletal material. On the windward side of the bank coral reefs develop.

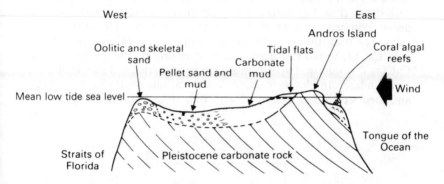

Fig. 4.7 The distribution of recent carbonate sediments on the Bahama Platform. The Straits of Florida act as a sediment trap for debris weathered from the Florida hinterland.

Modern carbonate producing environments show many subtle and important variations in grain type with depositional environment which are too complex for the present discussion. However, it is worth emphasising that the fundamental criterion for the development of a carbonate platform is the exclusion of silicate debris from continental weathering. Since the transport of such debris across the oceanic depths and up the side of the platform is impossible, an isolated platform such as the Bahama Bank can support prolific productivity. It is, therefore, an effective site of $CaCO_3$ removal from the oceans. The area of prolific carbonate sediment production is approximately 9000 km^2 and the average rate of accumulation has been estimated as 50 mg cm^{-2} yr^{-1} by Broecker and Takahashi (1966).

Total Rate of carbonate production $= 9000 \times 10^{10} \times 50$
$$\times 10^{-3} \text{ g yr}^{-1}$$
$$= 400 \times 10^{10} \text{ g yr}^{-1}$$
$$= 4 \times 10^{10} \text{ mol yr}^{-1}$$

Thus this area on the Grand Bahama Bank, representing approximately 10^{-5} of the surface area of the oceans, removes approximately 0.5% of the Ca^{2+} added by rivers to the oceans.

Controls of Calcium Carbonate Dissolution in the Oceans

Calcium carbonate, as the skeletal remains of the foraminifera, is abundant in deep sea sediments at depths of 1–4 km. However, below 4 km the $CaCO_3$ content of the sediments decreases rapidly with depth (Fig. 4.8), reaching very small values below 6 km. The deeper sediments consist of clays coated with iron oxide, called red clays. The transition in sediment mineralogy from red clay mixed with $CaCO_3$ to red clay alone can be demonstrated to arise from $CaCO_3$ dissolution. Experiments have been carried out in which foraminifera samples were suspended at various depths in the ocean and their weight loss measured with time (Fig. 4.9). The observed

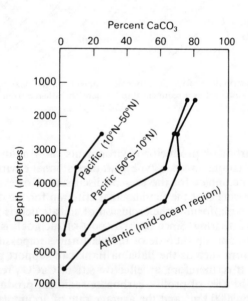

Fig. 4.8 Variation in $CaCO_3$ content with water depth for deep sea sediments. Most sediments from water depths of greater than 4 km show greatly reduced $CaCO_3$ contents (from Berner, 1971; by kind permission of McGraw Hill Book Company).

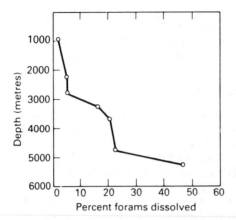

Fig. 4.9   The rate of dissolution of foraminifera over a period of 4 months at different water depths in the Pacific Ocean. Rapid rates of dissolution occur at water depths in excess of 4 km (from Berner, 1971; by kind permission of McGraw Hill Book Company)

and experimental data are closely comparable. The rate of $CaCO_3$ dissolution increases rapidly with depth, complete solution occurring below 4 to 5 km, at the carbonate compensation depth (CCD). The fact that no dissolution occurs in surface waters corresponds with the deduction that such waters are probably oversaturated with respect to $CaCO_3$. It seems reasonable to conclude that deep ocean water is undersaturated with $CaCO_3$ and that the CCD represents the boundary where sea water passes from oversaturation to undersaturation. This is, however, only partly true. Deep ocean waters do become undersaturated because of the effects of decreasing temperature and increasing pressure on the carbonate equilibria, and also because of the net production of CO_2 from the aerobic oxidation of organic matter (see Chapter 5). However, rate processes are important in determining the position of the CCD. The dissolution of carbonate minerals in slightly undersaturated water is slow, so that carbonate remains can accumulate if their rate of deposition is faster than their rate of dissolution. The observation that the CCD is not a simple depth, but is often a zone, suggests that a transition from oversaturation to undersaturation is not the only governing factor. Thus the influence of rate of carbonate production/dissolution indicates that the CCD provides another example of the effect of kinetics in environmental chemistry.

   The removal of Ca^{2+}, Mg^{2+} and HCO_3^- from the oceans by the formation of skeletal material is opposed by dissolution in deep waters. Removal processes predominate and the effect of skeletal

$CaCO_3$ precipitation and evaporite deposition together account for 61 % by weight of the annual discharge of dissolved solids by rivers. The major species remaining are HCO_3^-, SiO_2 and Mg^{2+} with smaller amounts of K^+ and Na^+ (Table 4.5).

Silica Formation in the Oceans

The two previous budget calculations have been able to use the major dissolved species in sea water to produce simple inorganic compounds which are common components of sedimentary rocks. A reasonable test of the quantitative predictions of these budgets is to compare the relative proportions of the different components produced in the budget with their abundancies in ancient sedimentary rocks. This argument rests on the hypothesis developed earlier that if ocean chemistry has been in a steady state for a substantial period of time, then the composition of ancient sediments formed to maintain a steady state in the past should be generally similar to those needed to maintain the present steady state. In this respect the predictions of carbonate and evaporite formation are broadly consistent with their observed abundance in ancient sedimentary rocks. Unfortunately the same is not true if all available SiO_2 was precipitated as silica in this stage of the budget.

At the present rate of flow of dissolved SiO_2 into the oceans from rivers, the volume of SiO_2 in present sediments would be much greater than that present in ancient sedimentary rocks. Dissolved silica in the oceans occurs in the form H_4SiO_4 and is precipitated as SiO_2:

$$H_4SiO_4 \text{ (aq)} \rightleftharpoons SiO_2 \text{ (s)} + 2H_2O$$

Precipitation of all 710×10^{10} moles H_4SiO_4 added annually by rivers would give an amount of silica, present as SiO_2, which would represent approximately 15 % of the volume of modern sediments. In fact, in both modern and ancient sediments silica as SiO_2 only constitutes between 1–5 % by volume. These observations suggest that a substantial proportion, but not all, of the dissolved silica added by rivers is precipitated as free SiO_2, i.e. not present in aluminosilicates, in ocean sediments. The behaviour of silica in the oceans has been the subject of much study, at least partly aimed at resolving this difficulty and some of the more important aspects of its chemical budget are discussed below.

In solution in ocean waters silica occurs almost entirely as H_4SiO_4. The first dissociation constant of H_4SiO_4 is $10^{-9.9}$, so for the dissociation

$$H_4SiO_4 \rightleftharpoons H_3SiO_4^- + H^+$$

$$\frac{[H_3SiO_4^-]}{[H_4SiO_4]} = \frac{10^{-9.9}}{[H^+]}$$

As the pH of seawater is approximately 8.1, the species H_4SiO_4 is nearly two orders of magnitude more abundant than $H_3SiO_4^-$. Silica is also an important nutrient for phytoplankton, especially diatoms, and is the major constituent of a number of solid phases found in the marine environment, e.g. aluminosilicates of terrestrial origin and solid forms of SiO_2 produced by biological activity or derived from land sources. However, the emphasis in this section will be on the behaviour of dissolved silica, although there is very considerable interaction between dissolved and particulate forms and the distinction cannot be applied at all rigorously.

Attempts to balance the silica budget by various investigators have identified a number of possible inputs and outputs, although quantifying these has proved difficult (Table 4.8). Not significant indicates that the magnitude of that flux is not considered to be important in terms of the total budget. A question mark means that the magnitude of the flux is unknown but may still be important.

It is apparent that there is considerable uncertainty concerning the magnitude of almost all of the terms in the budget. In these circumstances it is hardly surprising that it is impossible to say unequivocally whether or not the budget balances, i.e. whether the sums of the inputs and outputs are equal. In spite of this, in constructing the budget several useful things have been achieved. For example, it has been necessary to consider all of the processes which add and remove dissolved silicon to and from the oceans. In so doing lack of quantitative knowledge concerning many of them has been identified and attention has been focused on those processes which are worthy of further study. Other processes can apparently be ignored, at least in the budgeting context.

In terms of the present ocean budget, the detailed analysis in Table 4.8 indicates that at least 25–50 % of the river discharge of dissolved silica may be removed as skeletal remains into sediments. In order to allow continuation of the budget approach, the purely arbitrary value of 25 % is chosen for the removal of dissolved silica into sediments. Deposition of silica on this scale would represent approximately 4 % of the total volume of sedimentary rocks, that is within the observed range but towards its upper extremity. Therefore, 180×10^{10} moles of SiO_2 are removed annually into sediments and 530×10^{10} moles SiO_2 have to be carried through to the final stage of the budget.

Table 4.8 The silica budget of the oceans

	Flux estimate of silicon $(10^{14} \text{ g yr}^{-1})$
Inputs:	
River inflow	2.0
Dissolved silicon is one of the major elements in river water so that fresh water inflow constitutes an important source of silicon to the oceans. Some, probably less than 20%, of the total dissolved silicon in rivers may be removed from solution in the estuarine zone by biological and/or chemical processes.	
Undersea volcanic activity	Not significant
Weathering of suspended particles, bottom sediments and rocks	Not significant
Diffusion out of sediment pore waters	0.5 – 4.0
The interstitial water of marine sediments generally contains more dissolved silicon than the overlying sea water, so that the concentration gradient will drive a diffusive flux of silicon from sediment to water.	
Antarctic weathering	?
Erosion and weathering of the antarctic rock mass produces large amounts of fine particulate material, rock flour, and it has been suggested that on meeting sea water such material may readily dissolve so constituting a source of dissolved silicon to the oceans. Some recent evidence argues against the quantitative importance of this flux, so at present it is undecided.	
Total input	2.5 – 6.0 (+ ?)
Outputs:	
Deposition of oceanic Si-rich biogenic sediments	0.4 – 0.9
Some of the silicon fixed by plankton in the oceans escapes dissolution in the water column to form Si-rich oozes in certain oceanic areas, predominantly as a belt round Antarctica.	
Deposition of Si-containing biogenic sediments in coastal waters	?
The mechanism of this process is similar to that for the deposition of oceanic Si-rich biogenic sediments, except that the removal occurs in nearshore areas where the biogenic flux of silicon to the sediments is generally difficult to quantify against the high background flux of other particulate material into the sediments. However, this process may be an important output mechanism.	
Reaction of dissolved silicon with clay minerals in sediments	?
Reverse weathering (page 122) is a potential sink for dissolved silicon and is presumed to occur somewhere in the marine system, possibly in buried sediments	
Total output	0.4 – 0.9 (+ 2?)

Recycling of Silica in the Water Column

A considerable amount can be discovered concerning the behaviour of many elements in the marine environment by constructing mass balances of their known inputs and outputs to and from the oceans. However, although useful, such an exercise, in which the oceans are treated essentially as a black box, may hide important information about the way in which the elements are recycled within the sea water itself. Dissolved silica is an important nutrient element and as such shows pronounced internal cycling within the water.

Vertical profiles of dissolved silicon in the major oceans are shown in Fig. 4.10. It is apparent that there is a large vertical gradient in dissolved silicon. The low near-surface values are due to consumption of silicon by marine plankton, such as diatoms and radiolaria, which use this element in their skeletal structure. After death the plankton sink and may dissolve and decompose in the water. In the same way that the sedimentation of decaying organisms in lakes and rivers causes depth changes in water chemistry (page 85), so the breakdown of siliceous plankton leads to a general increase in dissolved silicon levels with depth in the sea-water column. Some of the sinking plankton escape complete dissolution and settle on the ocean bed to form the siliceous sediments (Fig. 4.11). The rather large difference in dissolved silicon concentrations in the deep waters of the Atlantic and Pacific Oceans is thought to be a result of the different water circulation patterns in the two oceans. The Atlantic is a source of deep water to the Pacific and only

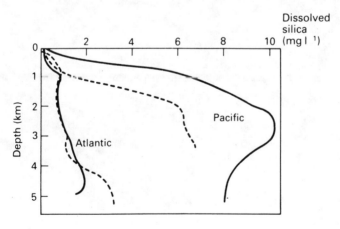

Fig. 4.10 Vertical profiles of dissolved silica in the Pacific and Atlantic Oceans. The Atlantic supplies deep, silica-rich water to the Pacific, but only receives silica-poor surface water in exchange (from Heath, 1974).

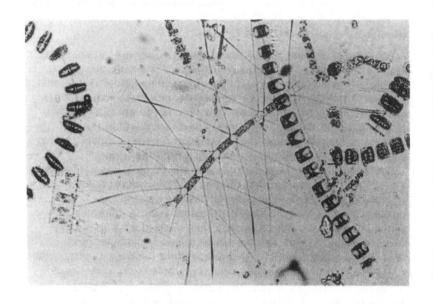

Fig. 4.11 Marine diatoms with siliceous skeletons. Field of view: a, 500 μm; b, 200 μm (courtesy of Lowestoft Fisheries Laboratory, Crown copyright).

gains Si-poor surface water in return. In contrast the Pacific gains Si-rich deep water from other oceans and looses only surface water from which biological activity has stripped most of the dissolved silicon.

Several estimates have been made of the yearly amount of dissolved silicon abstracted from surface waters by phytoplankton and the numbers range between 10^{16} and 10^{17} g of silicon per year. It is instructive to compare this internal cycling flux with the input/output fluxes given in the budget approach presented earlier. It is clear that internal processes turn over of the order of 100 times more silicon each year than the amounts entering or leaving the oceanic reservoir. This type of behaviour is typical of nutrient elements and is exhibited by none of the other oceanic dissolved species considered here. Although Ca^{2+} is also used in biogenic processes its rate of abstraction from surface waters is too small to produce vertical profiles of the type shown by silica.

Interactions between Particulate Material and Sea Water

After the third stage in the oceanic budget chemical sediments forming in the oceanic reactor vessel are capable of removing all the SO_4^{2-}, Ca^{2+} and Cl^- ions added by river discharge. The budget has also been significantly depleted in Na^+, Mg^{2+}, HCO_3^- and SiO_2 but still contains large proportions of these ions together with K^+ (Table 4.5). Removal of these components poses a problem as, apart from the phases already considered, there are no simple inorganic solids which are typically found in sediments. However, interactions between sea water and the particulates supplied by rivers have not yet been considered. Studies of the chemical behaviour of particulates entering the marine environment indicate that apart from ion exchange (page 74), the particles are probably inert to sea water. Ion exchange reactions are probably important in maintaining the oceanic balance of K^+ and Na^+, but are much less important for other cations. Comparisons of the composition of river clay and its changes on entering sea water indicate that both ions replace Ca^{2+}.

$$Ca - clay + 2K^+ \rightleftharpoons (K^+)_2 - clay + Ca^{2+}$$

Electroneutrality requirements demand that each mole of Ca^{2+} is replaced by 2 moles of K^+ or Na^+. Therefore, 189×10^{10} moles of K^+ added by rivers annually and removed by clays would generate 95×10^{10} moles of Ca^{2+}. The corresponding figures for Na^+ exchange generate 93×10^{10} moles of Ca^{2+}. The extra Ca^{2+} from both sources is precipitated as carbonate and consumes an additional 376×10^{10} moles of HCO_3^- (Table 4.5). An appropriate amount of

Mg^{2+} (page 108) will be precipitated together with this extra $CaCO_3$.

The ocean budget now needs removal processes for 500×10^{10} moles Mg^{2+}, 350×10^{10} moles SiO_2 and 958×10^{10} moles HCO_3^-. Studies of modern marine sediments suggest no other major interactions between sea water and particulate material, although structural changes in clay minerals do occur the deeper one penetrates into buried sediments. A key observation in this argument is provided by studies of ancient sedimentary rocks now exposed on the continents. Here the passage of time and the increased temperatures and pressures associated with deep burial have patently produced modifications of the original weathered silicates. It has been postulated that the reconstitution of clay minerals involves the reaction of weathered silicate debris from the continents with cations, SiO_2 and HCO_3^- to produce new clay minerals and sheet silicates:

$$\text{Degraded aluminosilicate} + HCO_3^- + \text{Cations} + SiO_2$$
$$\rightarrow \text{cation aluminosilicate} + CO_2 + H_2O$$

This type of reaction can be considered as reverse weathering since new sheet silicates are formed from weathered aluminosilicates plus dissolved constituents and CO_2 is released. The weathering reactions considered earlier are precisely the reverse of this, involving the consumption of CO_2 and production of HCO_3^-, together with the release of cations from the silicate structure. So far as the budget for the earth-air-water factory is concerned, the solid products of continental weathering can be crudely considered to have a kaolinitic composition, in so far as this mineral contains no metal cations and has a low Si/Al ratio. The following example is derived for the particular case of magnesium, as this is the major unused cation in the budget. The silicate transformation to produce chlorite is as follows:

$$Al_2Si_2O_5(OH)_4 + 5Mg^{2+} + SiO_2 + 10HCO_3^-$$
$$\rightarrow Mg_5Al_2Si_3O_{10}(OH)_5 + 10CO_2 + 3H_2O$$
Chlorite

If the mass balance reaction is written to use up all the remaining 500×10^{10} moles Mg^{2+}, chlorite formation will also consume 100×10^{10} moles SiO_2 and 1000×10^{10} moles HCO_3^- to produce 1000×10^{10} moles CO_2.

This particular example of a reverse weathering reaction leaves a budget now containing a significant surplus of SiO_2 and a slight deficiency of HCO_3^-. The surplus of SiO_2 is as expected, in view of the large uncertainties involved, whilst that for HCO_3^- is negligible by comparison to the total HCO_3^- input. The magnesium input and output can be considered as balanced, in relation to the errors and assumptions involved in these estimates.

Reverse weathering reactions provide an attractive solution to the problem of ocean balance but it must be stressed that there is no unequivocal evidence for their occurrence. It is possible instead that a magnesium balance in the oceans is maintained through the interaction of sea water with the basaltic rocks injected at the mid-ocean ridges between rifting continents. Experimental studies of basalt-sea water interactions are encouraging and further work is in progress on this question.

Anthropogenic Inputs to the Oceans

In this section the role of the oceans as a source or sink for a number of pollutants will be discussed. The pollutants chosen are both man-made (e.g. Freons, DDT and the PCBs) and natural, but increased by anthropogenic activity, (e.g. CO_2). For all the chemicals considered the main transport route to, or from, the marine environment is through the atmosphere, at least over the open oceans. In coastal waters the situation may be more complex since inputs in rivers discharging into the sea should also be quantified. For simplicity, river sources will be ignored here.

The Freons

Freons are a family of simple compounds having low molecular weights and containing the atoms C, Cl and F, and sometimes also H and Br. The two most important ones from the environmental standpoint are Freon-11 (CCl_3F) and Freon-12 (CCl_2F_2), whose structures are represented diagramatically in Fig. 4.12. As far as is known they are entirely man-made and are used in roughly equal

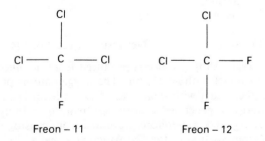

Fig. 4.12 The structure and composition of Freon-11 and Freon-12.

quantities as refrigerants and as propellants in aerosol spray cans. However, as might be expected, because refrigerants are used in closed systems, the use of Freons as propellants accounts for about 75 % of their emissions to the atmosphere. Total world production of these compounds has increased from virtually zero in the early 1940s to almost 10^9 kg yr^{-1} in 1974, although in the last few years production has levelled out at the 1974 figure or even decreased slightly, probably due to well-publicised criticism of their possible adverse effects on the environment and human health (see page 125).

The freons are remarkably inert under normal conditions and, in particular, their inflammability and their absence of toxicity are important with respect to their use as aerosol propellants in hair sprays, deodorants etc. Once released into the atmosphere the relatively inert behaviour of the Freons has led to the suggestion that they would be ideal tracers with which to study the dispersion of pollutants from urban sources. However, there do appear to be two routes by which freons are removed from the atmosphere, or transformed therein. One is by dissolution in the oceans and the other, by breakdown in the stratosphere. The Freon F–11 will be used to illustrate both processes.

The Stratospheric Sink   Under the action of solar ultraviolet radiation in the wavelength range 180–220 nm, the following sequence of reactions can be initiated by F–11 reaching the stratosphere from the troposphere:

$$CCl_3F \xrightarrow{uv} CCl_2F + Cl \tag{4.1}$$

F–11   atomic chlorine

$$\rightarrow Cl + O_3 \quad \rightarrow ClO + O_2 \tag{4.2}$$

ozone

$$ClO + O \rightarrow Cl + O_2 \tag{4.3}$$

atomic
oxygen

$$\therefore \quad O_3 + O \rightarrow 2O_2 \qquad \text{Net result of reactions (4.2) and (4.3)}$$

The atomic oxygen shown above is produced from the photolysis of O_2 by light of wavelength < 242 nm. The oxygen atoms produced in this way can then react with molecules of O_2 to form O_3, provided the excess energy so generated is absorbed by an inert body, usually a molecule of N_2, and is therefore prevented from causing the immediate decomposition of the O_3. Atomic oxygen and ozone are, therefore, both present as natural constituents in the stratosphere.

The net result of the reaction scheme illustrated in Equations (4.1)–(4.3) is to bring about the destruction of F–11 (Equation (4.1)) and O_3 via the chain reactions (4.2) and (4.3). It is apparent that this ozone destruction mechanism is initiated by the atom of chlorine produced by uv irradiation of the F–11 molecule and that, since in (4.3) one atom of chlorine is produced for each one consumed in (4.2), breakdown of each F–11 can lead to the destruction of many ozone molecules. However, the process is not quite as efficient as indicated by the above scheme since some of the chlorine atoms are removed from the cycle by reactions with molecules of methane (CH_4) gas to form HCl, which is eventually carried down into the troposphere by rain.

Although the reaction scheme outlined above is a considerable oversimplification of the chemistry of ozone in the stratosphere and the effect of compounds like Freons upon it, it does show how the stratosphere can act as a sink for such compounds. However, the real importance of Freon destruction in the stratosphere is that it could lead to a reduction in the amount of O_3 in this region of the atmosphere. Ozone is a powerful natural absorber of high energy (short wavelength < 290 nm) uv radiation from the sun, which is the reason why photolytic breakdown of Freons does not occur to any extent in the troposphere. Therefore, if stratospheric O_3 levels decrease, more short wavelength radiation will reach the earth's surface. High energy uv radiation can cause sunburn and it is thought that prolonged exposure may lead to an increased incidence of skin cancer. It is very difficult to predict with any certainty the magnitude of the ozone depletion to be expected from present and possible future use of Freons but current estimates indicate a mean figure of 7% ozone depletion, with uncertainty limits of at least 2 to 20%. The average figure corresponds to a yearly consumption in the stratosphere of about 2% of the amount of F–11 at present in the atmosphere. Even less certain than the ozone depletion estimate is whether a stratospheric O_3 decrease of this magnitude can have any significant effect on the occurrence of skin cancer, since uv radiation appears to be only one of a number of factors determining the incidence of this disease.

The Oceanic Sink In order to establish whether the oceans are a source or sink for any gas it is necessary to know if the surface water is over or under-saturated with respect to the atmospheric concentration of that gas. When the water is under-saturated this implies a flux (F) of the gas from air to sea and vice versa for supersaturation. The magnitude of the flux is proportional to the degree of under or over-saturation:

$$F \propto \Delta C \tag{4.4}$$

where ΔC is the concentration difference between air and sea, which drives a flux across the interface. Equation (4.4) can be written

$$F = k\Delta C \qquad (4.5)$$

where k is a constant of proportionality, called the transfer velocity, which quantifies the rate at which a gas crosses the sea surface. The value of k varies with the degree of mixing at the interface, i.e. the rate of transfer is faster in turbulent as compared to calm conditions, although its relationship to wind speed, for example, is often not well-established. The transfer velocity also varies between gases, but two groups can generally be identified with the variation in k within each group being quite small. The Freons and CO_2, as well as other relatively insoluble and/or chemically unreactive gases in water, e.g. N_2, O_2 and Rn, all fall into the group for which the main barrier or resistance to transfer across the interface is presented by the liquid water close to the surface. For gases of high solubility and/or chemically reactive in water, e.g., SO_2, NH_3, and H_2O itself, the main resistance to transfer is from the air on the gas side of the interface. DDT and the PCBs, are rare examples of compounds whose gaseous forms seem to fall between the two major groups and whose air-sea transfer is probably affected by resistance in both the liquid and gaseous phases on either side of the interface.

Equation (4.5) shows that the value of F for any gas can be calculated if the interfacial concentration difference and the appropriate value of the transfer velocity is known. For Freon–11, whose transfer will be liquid phase controlled, like CO_2 and Rn, the globally averaged value of k can be obtained by several techniques. Firstly, the distribution of $^{14}CO_2$, both natural and produced from the detonation of nuclear devices, between the atmosphere and the oceans can be measured, and secondly the concentration of a radioactive isotope of radon can be measured in near-surface sea waters. These suggest $k \simeq 15\ \mathrm{cm\ h}^{-1}$ for the group of liquid phase controlled gases, which includes the Freons. In principle it is straightforward to obtain values for ΔC since all that is required are measurements of the concentrations of the gases concerned in surface sea water and the overlying air mass. In practice, measurement is often very difficult for two reasons. Firstly many of the gases of interest occur at very low concentrations in the environment, e.g. in the marine atmosphere there is approximately 1 part of F–11 for every 10^{10} parts of air. Secondly, these gases also show substantial temporal and spatial variability so that large numbers of measurements must be made in order to obtain realistic average air and water concentrations. The most extensive and consistent set of data for F–11 (Fig. 4.13) is that obtained by Lovelock *et al.* (1973) who measured air and near-surface water

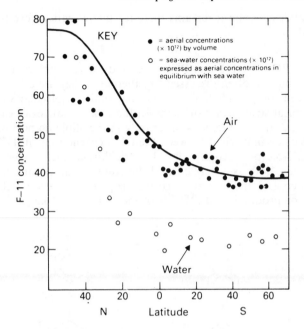

Fig. 4.13 The distribution of Freon-11 in and over the North and South Atlantic Ocean. Sea water appears undersaturated with respect to the overlying atmospheric concentrations of Freon-11, suggesting a flux from the atmosphere to the sea (from Lovelock *et al.*, 1973).

concentrations on a cruise of the research ship Shackleton in the North and South Atlantic. Amounts in the water are expressed as equivalent air concentrations and show that at almost every sampling position the water is undersaturated with respect to the atmospheric concentration, thus indicating a flux of F–11 from air to sea. Using the measured concentration difference from Fig. 4.13 and the value of $k = 15$ cm h^{-1} the global air to sea flux of F–11 is approximately 7×10^6 kg yr^{-1}. There are large assumptions behind this calculation, e.g. the k and ΔC values used may not be truly representative averages over the whole of the ocean surface at all seasons. Despite such extrapolations the calculated flux is almost certainly of the right order and indicates that the uptake of F–11 by the oceans each year removes slightly less than 1 % of the amount in the atmosphere.

Comparison of the stratospheric and oceanic sinks for F–11, given above, shows them both to be small and of roughly equal magnitude, i.e. about 2 % and 1 % yr^{-1}, respectively. The small percentage removal of the two known sinks leads to F–11 having a relatively long atmospheric residence time, i.e. 25–30 years, in the atmosphere and so after emission it remains there for many years before being removed.

DDT and PCBs   DDT and PCBs are used for very different purposes but are often discussed together since they are all chlorinated compounds having somewhat similar molecular weights and chemical structures. For example, the representatives of each family illustrated in Fig. 4.14 differ only in the aliphatic portion of the molecule sandwiched between the two benzene rings in DDT. DDT (2, 2-bis (p-chlorophenyl)-1, 1, 1-trichloroethane) is widely used as a biocide, both as a public health measure to kill mosquitoes in malarial control and to protect cotton crops from insect attack. Worldwide production of DDT is of the order of 10^8 kg yr^{-1}.

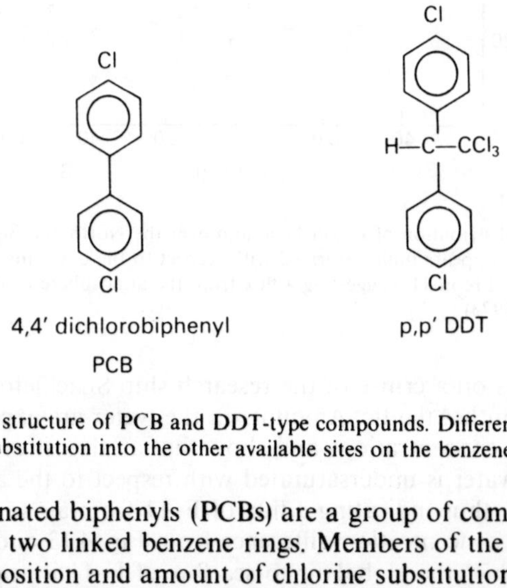

4,4' dichlorobiphenyl
PCB

p,p' DDT

Fig. 4.14   The structure of PCB and DDT-type compounds. Different PCBs may be formed by Cl substitution into the other available sites on the benzene rings.

Polychlorinated biphenyls (PCBs) are a group of compounds formed from two linked benzene rings. Members of the group differ only in the position and amount of chlorine substitution in the benzene rings. PCBs are very stable, inflammable, of low volatility and high dielectric constant. Total worldwide production is about 5×10^7 kg yr^{-1} of which much is used as dielectrics in transformers and capacitors, and as plasticisers, hydraulic fluids and lubricants.

Much of the DDT and a significant fraction of the PCBs is used in such a way that it readily enters the environment, e.g., spraying and dusting of crops with DDT. Both these compounds have been found all over the earth, even in regions such as the Poles which are far from their sites of application or use. It has been argued that transport via the atmosphere is probably the only route by which such a wide dispersion could occur in the relatively short time for which these compounds have been available commercially (i.e. post-1945). In this situation, the question arises as to what fraction of these compounds is removed from the atmosphere to the oceans by transfer across the air-sea interface.

Unfortunately calculation of the size of the oceanic sink is considerably more complex than for the Freons. For F-11 it was only necessary to consider input to the oceans from the gas phase since no significant amount of this compound appears to exist in the atmosphere in any other form, i.e. as solid particles or dissolved in water drops. For DDT and the PCBs transport in the atmosphere occurs in solid, liquid and gas phases and each must be considered to obtain the total flux into the oceans. Furthermore the behaviour of the gaseous forms of DDT and PCBs is less well-understood than for the Freons and it has not yet been established whether gas or liquid phase resistance, or a mixture of both, controls transfer of the gases across the air-sea interface. Therefore, there is uncertainty concerning the value to be used for the transfer velocity. In addition, analysis of these compounds at the very low concentrations at which they occur in the marine environment is extremely difficult, with the consequence that the data base for use in the flux calculations is very small. The best concentration data is for the western North Atlantic and Table 4.9 gives the results of flux calculations for this area of the oceans made by Bidleman *et al.* (1976).

Table 4.9 Fluxes of PCBs and DDT via gaseous, liquid and particulate routes to the western North Atlantic (from Bidleman *et al.*, 1976)

	Fluxes (g yr^{-1})			
	Gaseous	Liquid	Particulate	Total
PCBs	6×10^7 (42 %)	3×10^7 /21 %)	5.2×10^7 (37 %)	1.4×10^8
DDT	6×10^6 (42 %)	3×10^6 (21 %)	5.2×10^6 (37 %)	1.4×10^7

Note. Figures in brackets are percentages of the total.

The results indicate that all three modes of transfer are important, although gaseous and particulate fluxes predominate over input in rain. The total flux of PCBs is about ten times that for DDT despite the fact that the rate of manufacture is about 2:1 in favour of DDT. This probably arises from differences in their chemical reactivity since DDT is readily degraded by biological and/or chemical processes, whereas the high stability of PCBs leads to little if any breakdown of these compounds in the environment. Therefore, a much higher fraction of the PCBs released into the environment will reach the open oceans than is the case for DDT.

If the results given in Table 4.9 are extrapolated to estimate the amounts of these compounds taken up from the atmosphere by the oceans then approximately 5×10^6 kg yr^{-1} of PCBs and 5×10^5 kg yr^{-1} of DDT, or about 10% and 0.5% respectively of the total annual production, enters sea water in this way. However, such extrapolation is likely to overestimate the total flux since DDT and PCB measurements were obtained from oceans in the northern hemisphere, which are probably in contact with higher atmospheric concentrations than is the case in the southern hemisphere.

Carbon Dioxide

The environmental chemistry of CO_2 has received considerable coverage in other sections of the book so that here attention will be focused only on its transfer across the air-sea interface. As with the freons, only CO_2 in the gas phase need be considered since other modes of exchange are insignificant.

Although CO_2 is more soluble in water than, for example, O_2 or N_2, and does react with the water to form carbonic acid, it can be shown that neither of these factors is of sufficient magnitude to affect CO_2 transfer across the air-sea interface, which is liquid phase controlled. Since the transfer velocity for CO_2 is reasonably well known it should in principle be possible to calculate the flux of CO_2 into the oceans from the observed state of saturation of the surface sea water with respect to atmospheric CO_2 levels. However, measurements of CO_2 partial pressures in ocean surface waters indicate that the state of saturation is spatially rather variable, with some areas supersaturated and others undersaturated. No doubt this situation also exhibits substantial temporal variability. For these reasons a vast number of data points would be needed in order to obtain the net flux from the difference between the sums of the individual air to sea and sea to air fluxes. Therefore, for CO_2, the air to sea fluxes are generally obtained from box models of the atmosphere-ocean system (see page 34).

5 Formation of Mineral Resources in Sediments

Much of the river borne dissolved load added to the oceans is removed through the formation of chemically-precipitated sediments in different oceanic environments. These sediments may remain in suspension for considerable periods, depending on their mass and on the water conditions. For a typical density of $2.7 \, \mathrm{g \, cm^{-3}}$, approximately that of $CaCO_3$, SiO_2 and the aluminosilicates, particles classified in the sand size range settle out rapidly but clay-sized particles may remain in suspension for many months (Table 5.1). These sedimentation rates are derived assuming still water, but the effect of currents and turbulence significantly retards settling rates, such that fine-grained clay material may remain in suspension almost indefinitely. In these circumstances, currents may induce the selective sedimentation of different grain sizes, with the coarsest particles being transported only short distances from their source before settling out. This effect is strikingly illustrated by the fate of river-borne particulates.

The reduction of river current velocity in an estuary, where mixing with sea water first occurs, commonly results in extensive sedimentation to produce a widespread area of muddy tidal flats. The effect of reduced current velocity is enhanced by the change in salinity on passing from river water (average salinity $0.2\%_{oo}$) to sea water ($35\%_{oo}$). In the discussion of ion exchange (page 75), it was shown that clay minerals possess a residual surface negative charge which is mainly, but not completely, balanced by cations. Unless the negative charge is completely neutralised, clay minerals tend to remain in suspension as sedimentation involves grains coming into close proximity where like charges repel. The effect of surface charge is much diminished in sea water, where the higher concentrations of cations in solution allows the development of a mobile layer of cations in the fluid adjacent to the grain surface (Fig. 5.1). This phenomenum is called the electric double layer and the overall neutrality of the fixed layer, on the grain surface, and the mobile layer allows grains to approach each other and be deposited. This process is sometimes described as salting out particles from suspension in a fluid.

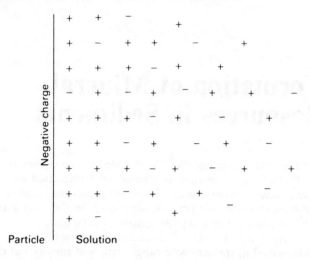

Fig. 5.1. The electrical double layer which comprises a fixed layer of charge on the particle and a mobile layer in solution. The latter exists because positive ions in solution are attracted to the particle surface.

Table 5.1 Sedimentation rates of different size particles

	Particle diameter (μm)	Time to settle 10 cm in still water
Clay	1	31h 7min
	2	7h 5min
Silt	5	1h 8min
	10	17min 30s
	20	4min 15s
Sand	1000	10s

The effects of current velocity and salinity greatly reduce the particulate load of a river and only the finer-grained material survives to enter the marine, rather than the estuarine, environment. Still further reductions in current velocity offshore lead to a characteristic pattern (Fig. 5.2) of decreasing grain-size with increasing distance from the land source. Clearly only the fine-grained clay fraction of river-borne particulates will be transported to the farthest parts of the ocean basins.

Modern oceanic sediments, therefore, consist predominantly of the fine-grained fraction of aluminosilicate material originally derived by continental weathering and subsequently transported by rivers and

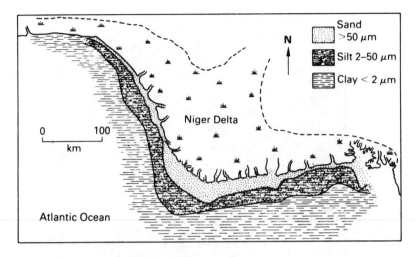

Fig. 5.2 Particle size distribution in bottom sediments on the Nigerian continental shelf. Grain size decreases away from the coast (after Allen, 1964).

ocean currents. Chemically-precipitated components on average constitute only a small fraction of modern oceanic sediments, with a still smaller proportion being composed of organic matter. Although the average composition is dominated by aluminosilicates, the range of compositions exhibited by modern oceanic sediments extends from pure chemical precipitates through intermediate mixtures to pure aluminosilicate debris. Such wide-ranging compositional variations result from differences in the relative rates of supply of chemically precipitated components, as compared to aluminosilicate debris, at different points within the ocean basins. Fig. 5.3. shows the distribution of the most abundant types of deep-sea sediment, i.e. aluminosilicate debris, followed by $CaCO_3$ and then SiO_2. The next most abundant sediment type are the minerals containing high concentration of manganese and iron, of which the manganese nodules are the most dramatic and frequently discussed examples.

Manganese Nodules

Manganese nodules were first discovered in the late nineteenth century, and are now known to occur on or near the sediment surface in virtually all deep-sea areas as well as in coastal waters and some fresh-water lakes. Typically manganese nodules are dull black, ovoid bodies (Fig. 5.4), roughly fist-sized, and weighing in the region of 1 kg, although very large nodules of up to 850 kg have been

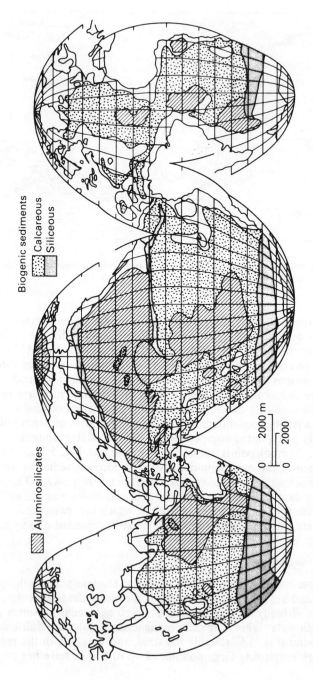

Biogenic sediments
Calcareous
Siliceous

Aluminosilicates

0   2000 m
0   2000

Fig. 5.3 The distribution of deep-sea sediments. Aluminosilicates, mainly derived by weathering, are distinguished from calcareous and siliceous biogenic components (after Riley and Chester, 1971; by kind permission of Academic Press, London).

Fig. 5.4 A field of closely-spaced manganese nodules on the bottom of the Antarctic Ocean. Mean nodule diameter is 6 cm (courtesy of C. D. Hollister).

dredged up from the sea floor. Mineralogically the nodules consist of a mixture of fine-grained oxides of iron (Fe_2O_3) and manganese (MnO_2), which may be associated with small amounts of clays, $CaCO_3$, SiO_2 and organic matter. A typical nodule contains about 20 % Mn, 20 % Fe and 1–3 % Ni + Cu + Co, by weight, but individual nodules can have compositions which differ from this by more than a factor of two.

It is, however, the presence of trace elements, especially those which are in short supply from land sources, e.g. Ni and Cu, which is responsible for much of the interest shown in manganese nodules in recent years. At the present time it is thought to be marginally economic to harvest nodules from the sea floor for the trace metals they contain and several companies are currently engaged in pilot mining schemes to test the practicability of this idea. Technical or other problems may mean that it remains cheaper to exploit land-based sources of these metals, although no doubt it will not be long before mining of deep-sea nodules becomes economically viable.

If manganese nodules are cut in half and the internal structure examined, it is generally found that the nodule consists of a nucleus of foreign material, often consolidated clay particles, volcanic glass,

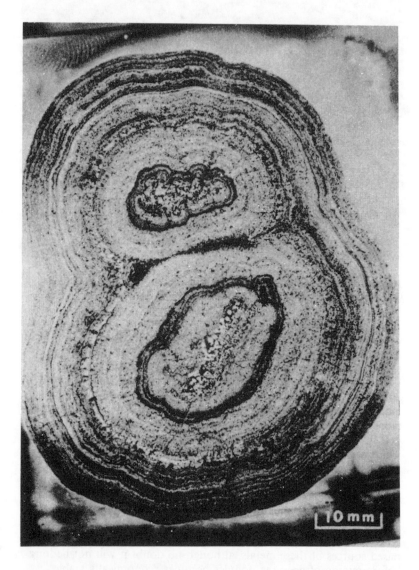

Fig. 5.5 Cross-sectional cut through a nodule with two growth centres. The altered volcanic material at the centres nucleated nodule growth. Crude growth rings represent temporal changes in composition and texture. The nodule has a diameter of 4 cm (courtesy of R. K. Sorem and E. R. Foster).

igneous rock fragments or a sharks tooth, surrounded by concentric light and dark bands of ferromanganese material. Fig. 5.5 shows a particularly fine example of a nodule with two growth centres, in this case altered volcanic material. From their appearance in cross-section nodules seem to grow by deposition of ferromanganese minerals onto a foreign nucleus, with subsequent addition onto the pre-existing outermost layer of deposited material. The somewhat different colour of adjacent layers, which makes the concentric growth pattern visible, is thought to be due to differences in their Fe/Mn ratios.

Manganese nodules from different marine areas appear to grow at very different rates. In coastal waters the rate of growth seems to be very fast since ferromanganese coatings several centimetres thick have been found on modern objects, such as automobile sparking plugs and 20th century naval shells dredged from the sea floor. However, most measurements of the age of various layers of deep-sea nodules indicate that here the growth rate is very slow being of the order of a few millimetres per million years, although not all researchers accept such low values. Assuming they are correct, the very slow rates of growth present a problem since the deep-sea sediment surrounding the nodule accumulates about a thousand times faster than the nodule itself, so prompting the obvious question of why the nodule is not rapidly buried by the much more rapidly accumulating sediment on which it sits. Various mechanisms have been proposed to overcome this problem. For example, since nodules are common in areas of strong bottom water currents it has been suggested that water movement may be an important factor in removing sediment from around the growing nodule. It has also been proposed that burrowing organisms and rat-tailed fish may also help to keep nodules at the sediment surface. However, these are no more than possible answers to the problem and it is difficult to know which, if any, of them, are actually operative.

The precise mechanism by which manganese, iron and other metals are incorporated into nodules is not well-established. However, it is generally assumed that deposition of manganese, and iron, is the result of a precipitation reaction caused by a change in the oxidation state of the element. Manganese dissolved in sea water has an oxidation number of two. If the Mn(II) is oxidised to the (IV) state then its solubility in the water is greatly decreased, so that it precipitates from the water. Therefore, at the surface of the growing manganese nodule, Mn(II) is thought to be oxidised to Mn(IV) and so precipitated onto the nucleus or outermost layer of ferromanganese mineral. At the same time other trace metals in the water, e.g., Ni, Cu, Co, coprecipitate along with the iron and manganese and/or are scavenged from the water by the highly absorbent freshly precipitated iron and manganese oxide surface, and are incorporated into the nodule. A possible role of biological processes, via attached bacteria or plankton, in the formation of manganese nodules has been suggested and is currently the subject of considerable research.

Whatever the exact mechanism by which dissolved manganese is removed from the water and precipitated as $Mn(IV)$ in the nodule, there is still the problem of identifying the source of the $Mn(II)$ in the water. There are thought to be three main sources of dissolved manganese in sea water, i.e. river inflow, submarine volcanicity and diffusion out of sediment pore waters. Rivers bring substantial amounts of dissolved and particulate manganese into the oceans and are obvious sources for the manganese found in rapidly accumulating near-shore nodules. Submarine volcanicity injects a variety of metals such as Mn, Ni, Co, Cu and Fe into sea water, mainly at the mid-ocean rift zones associated with sea floor spreading, and may be the source of these metals in nodules found on the surface of the flanks of the mid-ocean spreading centres. Measurements in the pore waters of some marine sediments show that they can contain higher amounts of dissolved manganese than the overlying sea water. These elevated concentrations arise from the conversion of particulate $Mn (IV)$ to soluble $Mn(II)$ due to the lower redox potential in the sediments arising from breakdown of organic material. The gradient of dissolved manganese between the sediment pore water and the overlying sea water implies a flux of dissolved manganese out of the sediments and a possible third source of the elements for nodule formation. Although some nodules appear to have fairly obvious sources for the manganese they contain, this is not a necessary prerequisite for their existence since much of the manganese input will become well-mixed in the oceans and is then available for uptake into nodules for which there is no obvious source of direct supply.

Microbiological Processes in Sediments

Individual components in oceanic bottom sediment mixtures may show little tendency to interact with sea water, without necessarily being at equilibrium. The reasons for this are usually kinetic, i.e. although reaction is energetically favourable the rate at which it occurs is very slow. One example of a reaction being kinetically inhibited may be that of reverse weathering, where weathered aluminosilicate debris from the continents are postulated to react with dissolved carbonates, silica and cations to produce new clay minerals and lattice silicates (page 122). An extensive study of modern surface environments has failed to show unequivocal evidence that this reaction occurs. Apart from cation exchange reactions, the aluminosilicate debris supplied by continental weathering appears to be chemically inert to sea water under low temperatures and pressures. Possibly any reaction is too slow to be perceptible. Certainly extensive alteration of aluminosilicates occurs at burial depths of a thousand metres or more, where increased temperatures and pressures produce more favourable reaction kinetics. Mixtures of the inorganic components common in sediments; the chemical precipitates, aluminosilicates, and sea water, may be thermodynamically unstable but kinetic factors effectively prevent reaction in near-surface sediments. Reactions involving organic matter, however, provide a striking contrast to the purely inorganic system.

In any sediment containing organic matter all the important reactions involve the oxidation of organic matter by inorganic components, aided by the catalytic action of different microorganisms. These reactions release significant amounts of energy which are used by the microorganisms for their respiration. Some such microbiological reactions, and their role in transforming sedimentary organic matter into coal and oil, are discussed below.

The organic matter incorporated into sediments may have many origins, representing the partially decayed residues of a variety of land and aquatic plants and animals. Together these sources provide a huge variety of different organic compounds, capable of interacting with each other and various inorganic components in an almost infinite number of ways. It is scarcely surprising that there is no detailed understanding of the reaction mechanisms by which organic matter is transformed into coal or oil, especially as the final stages of the transformation occur at burial depths which are relatively inaccessible. It is, however, possible to isolate several types of process which have an important influence on sedimentary organic matter in general, and which assist in the preparation of precursor petroleum or coal materials.

Living matter is thermodynamically unstable and its continued existence is possible only if energy can be derived from an outside source. Directly or indirectly this source is always solar radiation. Plants use solar radiation to overcome the unfavourable energy requirements of photosynthesis, which produces new cell material:

$$6CO_2 + 6H_2O \xrightarrow[\text{Solar radiation}]{\text{Chlorophyll}} C_6H_{12}O_6 + 6O_2 \qquad + 2879 \text{ kJ}$$

Animals use the reverse reaction, and, by consuming plant material, generate the energy they require. In both cases an energy input provides the means to generate new cell material for growth and to compensate for that lost by damage and decay.

Once life ceases, decay processes must prevail and may occur in a variety of ways, depending mainly on the availability of oxidising agents. Most decay processes are carried out by microorganisms, which play an essentially catalytic role. Under sterile conditions in the presence of air, the simple oxidation of organic matter by molecular oxygen is slow. The practice of deep-freezing and refrigeration depends on the inhibition of microbiological decay processes at low temperatures, enabling food (organic matter) to be preserved.

Decomposition in the presence of air is known as aerobic metabolism and is essentially the reverse of photosynthesis:

$$C_6H_{12}O_6 + O_2 \xrightarrow{\text{Microorganisms}} 6O_2 + 6H_2O \qquad -2879 \text{ kJ}$$

The formula $C_6H_{12}O_6$ represents the composition of carbohydrates $(C_nH_{2n}O_n)$, which are commonly found in sedimentary organic matter. For example, the compound cellulose comprises approximately half the cell-wall material of wood and other plants, and is a high molecular weight polymer built up from many carbohydrate units. Although there are a variety of different organic compounds present in sedimentary organic matter, an equation written in terms of the $C_6H_{12}O_6$ unit provides a schematic representation of aerobic metabolism, indicating that the major products are carbon dioxide and water. For many organic compounds complete breakdown to carbon dioxide and water may be impossible and in these circumstances reaction continues until the original organic matter has all been transformed to a partially degraded residue, which is inert to further microbiological attack:

$$RC_nH_{2n-1}O_n + O_2 \rightarrow RC_{n-1}H_{2n-3}O_{n-1} + CO_2 + H_2O$$

Organic   Partially
matter   degraded
                          residue

In the absence of oxygen, i.e. under anaerobic conditions, a different group of microorganisms use dissolved sulphate ions from sea water as an oxidising agent for sedimentary organic matter. This process is known as sulphate reduction. Using a simple carbohydrate it can be shown that both dissolved carbon dioxide, as bicarbonate ions, and hydrogen sulphide are produced:

$$C_6H_{12}O_6 + 3SO_4^{2-} \rightarrow 6HCO_3^- + 3H_2S \qquad -921 \text{ kJ}$$

Since many sulphides are insoluble, sulphide reduction characteristically results in the formation of metallic sulphides, in particular the mineral pyrite (FeS_2). The first step in pyrite formation is the generation of H_2S by microbiological sulphate reduction followed by the reduction of iron oxides present in the sediment and their conversion firstly to an iron monosulphide FeS. The transformation to FeS_2 is believed to involve the slow reaction of FeS with elemental sulphur. The various steps in the reaction are summarised below:

$$2Fe(OH)_3 + 3H_2S \rightarrow 2FeS + S + 3H_2O$$
$$FeS + S \qquad\quad \rightarrow FeS_2$$

The third microbiological process which is important in transforming sedimentary organic matter is fermentation, which differs from aerobic metabolism and sulphate reduction in being independent of an external oxidising agent. Microbiological fermentation reactions result in the oxidation of organic matter by the oxygen contained within the organic carbon compounds:

$$C_6H_{12}O_6 \rightarrow 3CO_2 + 3CH_4 \qquad\qquad -423\,kJ$$

An oxidised carbon species (CO_2) and a reduced species (CH_4) are generated simultaneously.

Each of these three microbiological processes is similar in that organic matter is oxidised to either CO_2 or dissolved carbonate species, but differ both in the nature of the accompanying products and in the energy yields. Aerobic metabolism generates no other important products apart from CO_2, but sulphate reduction results in pyrite formation and fermentation produces methane gas. Observations of modern organic-rich sediments indicate that the three microbiological processes occur sequentially. Aerobic metabolism is restricted to the top few centimetres and is followed by sulphate reduction, where sulphate is available, to depths of approximately 5m or less. Beyond this point fermentation reactions predominate. Several different factors combine to produce this sequential structure (Fig. 5.6). Firstly the order in which different microbiological processes occur represents the declining energy yield from the breakdown of each mole of organic matter. The greater the energy yield from a microbiological redox reaction, the more favoured are the microorganisms using that reaction for their respiration. Secondly, the depth to which each zone persists is controlled by the supply of oxidising agents, which are necessary to sustain that microorganism. Aerobic metabolism has the highest energy yield of the possible microbiological processes and is, therefore, the first dominant reaction. The zone thickness is determined by the depth to which dissolved molecular oxygen can diffuse from the overlying sea water. The thickness is small because the concentrations of dissolved oxygen in sea water are low ($< 2.5 \times 10^{-4}\,mol\,l^{-1}$) and because aerobic metabolism is the most rapid of all microbiological processes. The lower limit of aerobic

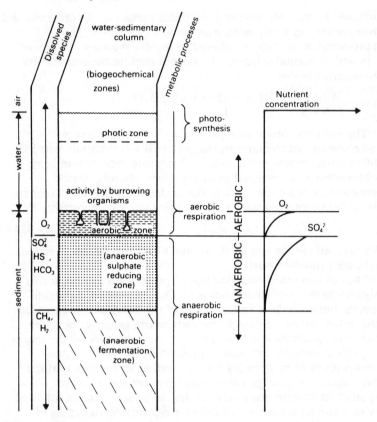

Fig. 5.6 An idealised vertical profile through a marine organic-rich sediment, showing different reducing zones which develop as a consequence of a microbiological succession (after Claypool and Kaplan, 1974).

metabolism is marked by the depth at which no dissolved oxygen can survive transport through the overlying sediment, without being consumed by aerobic metabolism. Below this depth anaerobic conditions prevail and the next energetically favourable reaction is sulphate reduction. The lower limit of the sulphate reduction zone is controlled by factors analogous to those defining the limit of aerobic metabolism, i.e. the depth to which dissolved sulphate ions can survive transport through the zone of sulphate reduction, before being completely consumed. This depth is larger than that for aerobic metabolism because of the larger concentration of sulphate in sea water (2.8×10^{-2} mol l^{-1}) and the fact that sulphate reduction occurs less rapidly than aerobic metabolism. Below the depth of diffusive sulphate penetration into the sediment fermentation, the least energetically favourable reaction, occurs to burial depths of approximately 1 km.

It can be seen from this that zone thickness depends on kinetic factors. Each successive microbiological process takes place only as long as the rate of supply of oxidising agent exceeds the rate of microbiological decomposition. When this constraint is removed the residual oxidising agent becomes rapidly depleted and that particular microbiological process is terminated to be replaced by its successor. The overall situation (see Fig. 5.6) is essentially static, with a series of microbiological zones developed at fixed distances below the sediment surface. However, in a natural system, sediment is continually being added to the surface although the effect of this does not necessarily alter the position of any given zone with respect to the sediment/water interface, since the factors which determine the extent of each zone are themselves depth-dependent. Therefore, the sequence and thickness of each microbiological zone may bear a constant depth relationship to the sediment/water interface, despite the addition of new material to the sediment surface. However, the addition of new sediment causes any particular layer of material to become further from the sediment/water interface and so to pass successively through each microbiological zone, as shown in Fig. 5.7

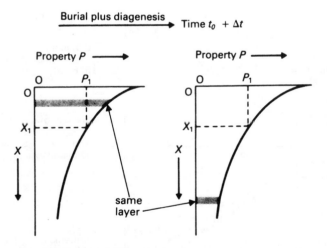

Fig. 5.7 Diagrammatic representation of an idealised relationship between burial depth and diagenesis. At a given depth x, the sediment has the property p, which does not change with time. The property p for the same layer of material does change with diagenesis as the sediment is buried to greater depths (from Berner, 1971; by kind permission of McGraw Hill Book Company).

The key factor which controls the evolution of organic rich sediments is the time spent in each successive zone, and this in turn is dependent on the rate at which new sediment is added to the surface. Consider the zonal sequence shown in Fig. 5.6 again. The sediment being deposited contains a fixed concentration of particulate organic matter which is at its maximum at the sediment surface. If deposition is absent then all the organic matter at the sediment surface is destroyed by aerobic metabolism and none will survive to be buried when sedimentation is resumed. Once the organic matter is destroyed microbiological processes cease and the surface sediment consists solely of the relatively unreactive inorganic components, probably aluminosilicate debris. However, if slow deposition occurs, the layers originally at the surface may pass through the zone of aerobic metabolism before all the organic matter is consumed. Sulphate reduction is now possible and the mineral pyrite is formed as a by-product. If deposition is rather more rapid, and the organic matter content sufficiently high, a proportion of organic matter may survive beyond the depth limit of sulphate reduction and fermentation reactions can occur. Thus the rate of sedimentation controls the duration of each successive microbiological process and consequently produces different types of sediment from the same initial mixture of organic and inorganic

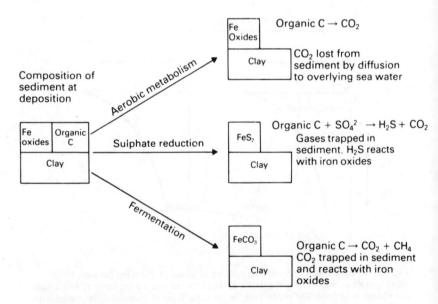

Fig. 5.8 Compositional variations in organic-rich sediments resulting from different types of microbiological activity.

components (Fig. 5.8). It should be noted that variations in the initial ratio of organic/inorganic components will to some extent balance the influence of rate of sedimentation. If the organic matter content of the sediment is high, it is more probable that some organic matter will survive burial to the lower zones even where the rate of deposition is slow. This zonal model of the burial changes in organic-rich sediments is highly idealised, not least because the thicknesses of each zone are dependent on many factors, but it does provide a useful framework for predicting the circumstances in which organic matter can become transformed to oil, gas and coal.

Formation of Petroleum

Petroleum is a complex and variable mixture of hydrocarbons, i.e. paraffins, cycloparaffins and aromatics[†], together with non-hydrocarbons. The structures of representative compounds from each group are shown in Fig. 5.9, together with their typical range of variability. These compounds may be separated on the basis of molecular weight by distillation, which is the simplest and most common method by which petroleum is refined. Each fraction (Table 5.2) has its own particular domestic or industrial usage. In passing it is worth noting that the sulphur-bearing compounds are responsible for the production of SO_2 (page 39) during the combustion of fuel oils.

The most striking feature about the organic compounds found in petroleum is the high proportion of low molecular weight compounds of a simple structure, compared to the prevalence of complex high molecular weight compounds in living organisms. If sedimentary organic matter derived from living organisms is the basic raw material for petroleum, then the effects of different source material, dispersal and depositional environments must be substantially offset by chemical changes after burial.

Whatever the source material, its subsequent evolution must follow a similar pathway in producing the small number of relatively simple organic compounds which constitute petroleum. It is, therefore, surprising that laboratory experiments with sedimentary organic matter do not produce petroleum-type mixtures of hydrocarbons under conditions that could conceivably exist in nature. Fatty acids, long-suspected to be the chief source of hydrocarbons, break down to form hydrocarbons and CO_2 at a measurable rate only when heated to temperatures of 300–400°C, whereas many petroleum-bearing sediments are unlikely to have reached temperatures greater than 100°C. Several different explanations have been proposed to explain this dilemma.

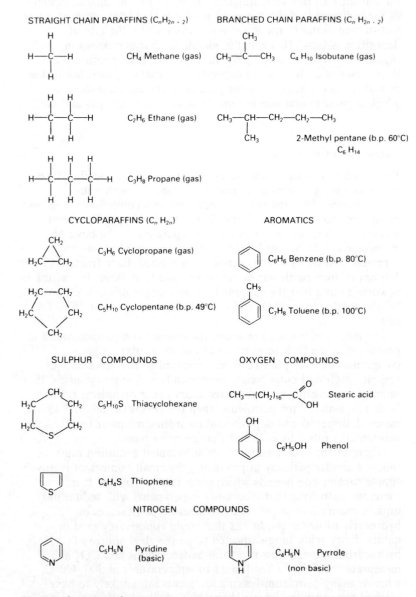

Fig. 5.9 Structures of some representative hydrocarbons and sulphur, oxygen and nitrogen-bearing organic compounds found in petroleum.

Table 5.2 Approximate compositions and boiling points of various petroleum fractions

Name of fraction	Representative hydrocarbons	Approximate boiling point (°C)
Natural gas	CH_4	-161
Liquefied gas (LP gas)	C_3H_8–C_4H_{10}	-44–$+1$
Petroleum ether	C_5H_{12}–C_6H_{14}	30–60
Aviation gasoline	C_5–C_9	32–150
Petrol	C_5–C_{12}	32–210
Naptha	C_7–C_{12}	100–200
Paraffin	C_{10}–C_{16}	177–290
Fuel oil	C_{12}–C_{18}	205–316
Lubricating oils	C_{15}–C_{24}	250–400

Firstly, hydrocarbons generally similar to those found in petroleum are present in small concentrations (< 0.3 mg l^{-1}) in living organisms. This has lead to the suggestion that petroleum pools represent accumulations of the hydrocarbons present in plants and animals, rather than the breakdown products of buried sedimentary organic matter. Such biogenic hydrocarbons can be detected in near-surface sediments and are in some ways strikingly similar to the hydrocarbons found in crude oils. Differences of detail exist however. Light hydrocarbons in the C_3 to C_{14} range are absent from modern organisms and near-surface sediments, but constitute approximately 50 % of an average crude oil. Furthermore, hydrocarbons in modern organisms and sediments show a preference for paraffins with a chain length composed of an odd number of carbon atoms, as compared to crude oils in which odd and even-numbered chains are present to a similar degree. These differences are difficult to explain if crude oils are predominantly biogenic accumulations and it seems more probable that a large proportion of the hydrocarbons are synthesised during burial.

Secondly, there are a number of ways in which the low temperature breakdown of organic matter may produce hydrocarbons at measurable rates. Microorganisms can produce methane by fermentation reactions at low temperatures and might conceivably be responsible for the production of other light hydrocarbons. However, the microbiological generation of hydrocarbons other than methane has never been demonstrated experimentally. Another possibility is that alpha-particles produced from radioactive decay bombard fatty acids, splitting off carboxyl groups to leave hydrocarbon residues. This process also results in the formation of helium and hydrogen, neither of which are present in sufficient abundance in petroleum or natural gas. Finally, it is possible that aluminosilicate minerals have a catalytic action on the thermal breakdown of organic matter to form hydrocarbons. Catalysis by aluminosilicates is certainly effective at high temperatures but cannot be unequivocally demonstrated at temperatures around 100°C.

Although the detailed mechanism by which hydrocarbons are produced is still not known the balance of evidence favours the slow, thermal degradation of organic matter by reactions such as decarboxylation, where carboxylic acids break down to lose their COOH groups leaving a hydrocarbon residue:

$$RCOOH \rightarrow RH + CO_2$$

In the zonal model of microbiological processes in organic-rich sediments, the lower limit of fermentation generally occurs at burial depths of around 1 km, at which point the average geothermal gradient[†] ($35°C \ km^{-1}$) has increased temperatures quite substantially. With further burial, temperatures are approached at which thermal decarboxylation of fatty acids and other processes may generate hydrocarbons.

Although the postulated temperatures are lower than those necessary for decarboxylation in the laboratory, reaction may be assisted by catalysis and may achieve an efficient transformation over the longer times available in natural geological systems. The zone of thermal decarboxylation extends over several kilometres, with the most intense generation occurring at depths of 2–3 km. The increasing temperatures associated with burial have other important effects on the crude oil: solids or low melting point liquids become more mobile; light hydrocarbons become more soluble in water; and large oil molecules may crack to form smaller molecules. These changes all assist the migration of crude oil from its source beds into the overlying reservoir rocks, e.g. porous sandstones, from which it may ultimately be extracted. Pressure effects are important in assisting the migration of oil from its site of generation to the reservoir, but are much less important than temperature in the actual process of hydrocarbon genesis.

A potential hydrocarbon source bed must, therefore, remain in the zone of thermal decarboxylation for a period of time long enough to allow efficient hydrocarbon generation and migration. If burial continues, the increased temperatures ultimately favour the formation of CH_4, which becomes increasing stabilised relative to other hydrocarbons at temperatures greater than 100°C (Fig. 5.10). The flow diagram presented in Fig. 5.11 summarises the different evolutionary pathways an organic-rich sediment may follow, depending essentially on its rate of burial. The best oil prospects result when a large proportion of the original organic matter

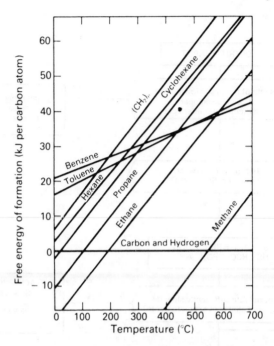

Fig. 5.10 The stability of hydrocarbons decreases with temperature but methane remains stable, indicated by a negative free energy of formation, to relatively high temperatures and may, therefore, be produced by the thermal degradation of other hydrocarbons.

survives passage through the zones of aerobic metabolism, sulphate reduction and fermentation. This indicates rapid burial to depths of approximately 1 km. On entering the zone of thermal decarboxylation burial rates should be slow enough to permit the efficient breakdown of organic matter to hydrocarbons. Where burial rates are more rapid the surviving organic matter breaks down to natural gas or methane, rather than more complex hydrocarbons, and the hydrocarbons produced earlier may have insufficient time to migrate to reservoirs and are themselves eventually degraded to methane. Gas reservoirs may, therefore, be produced from organic-rich sediments which have been buried rather too rapidly in their final stages.

Formation of Coal

Coal is a mixture of organic compounds of a high molecular weight and complex structure, containing a large percentage of carbon with

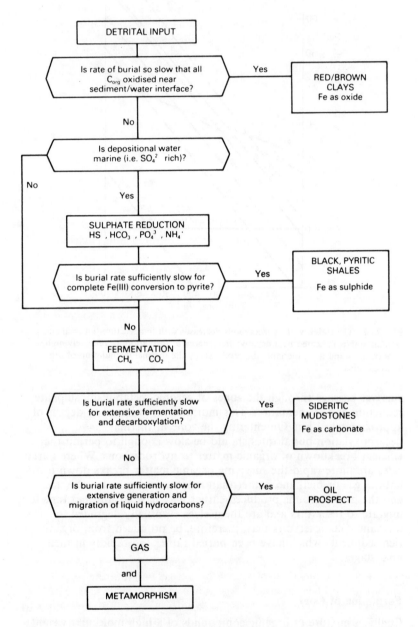

Fig. 5.11 Links between sediment mineralogy and rate of burial (after Curtis, 1977).

small amounts of hydrogen, oxygen and nitrogen. Traces of sulphur, which give rise to SO_2 on combustion, and phosphorus may also be present. In contrast to petroleum many of the organic compounds present in coal have not been identified, although it is clear that free carbon is not present except in coals which have been deeply buried and subjected to high temperatures.

There are other marked differences between petroleum and coal. Petroleum is a fluid which may have migrated large distances from its place of origin and thus does not retain fossils or structural features which indicate its origin. Coal betrays its origin by fossils and by its interbedding with sedimentary rocks and there is little doubt that it is a product of the partial anaerobic decomposition of buried terrestrial vegetation in a swamp environment. Coal is, therefore, derived primarily from land vegetation which is buried and transformed in a near-shore environment, whereas the precursor organic matter for oil probably originates from marine organisms, buried and transformed in an offshore marine environment. Despite recognition of this difference in environment of formation, the processes of coal formation are less well understood than those of hydrocarbon genesis. However, a reasonable picture of the processes involved, at the present state of knowledge, begins by distinguishing the different types of coal.

Two major types of coal can be recognised: humic coals and sapropelic coals. Humic coals develop from peat, which is an accumulation of dead vegetation at the site of the peat-forming plants. Sapropelic coals, on the other hand, are formed from organic-rich muds deposited in poorly-aerated lakes. Sapropelic coals pass through an aerobic decomposition process rather similar to that already described for petroleum. That this process results in coal rather than oil is believed to be due to the greater abundance of organic matter and its origin from land plants, rather than marine organisms. Sapropelic coals are relatively rare and only the humic coals will be discussed here.

Most humic coals originated from forest peats and, therefore, mainly from wood and bark substances, leaves and the roots of swamp vegetation. The major stages in the formation of coal are presented diagrammatically in Fig 5.12. The vegetation initially consists of carbohydrates, proteins and lignin the decomposition of which begins when the dead vegetation is incorporated into the soil system. Decomposition initially occurs under aerobic conditions (i.e. by the aerobic metabolism reaction on page 140) and the carbohydrates are rapidly broken down to CO_2 and water. The residues are thereby concentrated, and begin their transformation to humic substances, i.e. complex organic material with a high content of carboxylic and phenolic groups. The formation of humic

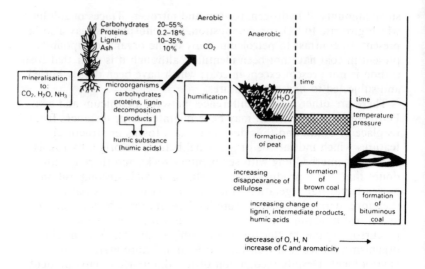

Fig. 5.12 Mineralisation and humification: the formation of peat, brown coal and bituminous coal.

substances is the most important process in peat development and is enhanced by the access of air, high temperatures and an alkaline environment. Continued decay under aerobic conditions is impossible in swampy environments where a change in water level may exclude contact with the atmosphere. When the water is depleted of oxygen, anaerobic conditions prevail and sulphate reduction may occur if the stagnant swamp waters were originally brackish. In sulphate-deficient freshwaters anaerobic conditions immediately give rise to fermentation reactions, as evidenced by the evolution of methane or marsh gas. During this stage in maturation important elemental changes occur in the composition of the decayed vegetation and its transformation products. Nitrogen and sulphur are much reduced as microbiological processes evolve ammonia and hydrogen sulphide. Up to this point the maturation processes are solely microbiological and result in the formation of peat, the average water content of which is approximately 90 %. Much of this water is adsorbed and can only be removed by drying as opposed to squeezing.

The processes which convert peat to coal are poorly understood, not least because of the difficulty in identifying the particular chemical compounds in coal. Transformation to coal only begins when the peat is covered by layers of sediment impervious to air, and can best be characterised by changes in the gross chemical composition. Time, temperature and pressure now become

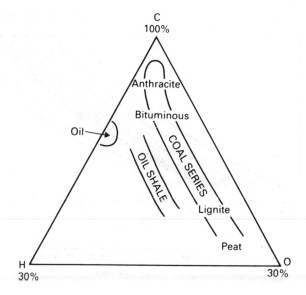

Fig. 5.13 Compositional relationships between coals, oil shale and crude oil (from Forsman and Hunt, 1959)

important factors and combine to increase the ratio of carbon to volatile constituents (O, H, N) in a sequence passing from peat to lignite (brown coal), through bituminous coals to anthracite (Fig 5.13). Changes in elemental composition are mirrored by a disappearance of recognisable plant remains, decrease in moisture content and an increase in hardness.

The position of a coal in this series is called its rank. Observations of many different coals show that rank generally increases with increasing depth of burial. In the past it was assumed that pressure was important in controlling rank, but laboratory experiments have shown that static pressure does not increase rank although dynamic pressure, i.e. pressure which changes constantly in magnitude and direction, may be of assistance. However, fluctuating pressures of this type are not common and it is now believed that the increase in rank with burial depth is primarily due to the increase in temperature caused by the geothermal gradient. An important correlation can be drawn between duration of heating, rock temperature and the rank of coal, as based on its volatile content (Fig 5.14). The same coal rank can be produced by either a short intensive heating or by low heating over long periods of time. Since temperature acts essentially by increasing reaction rates, e.g. a 10°C increase in temperature doubles reaction rate, this diagram

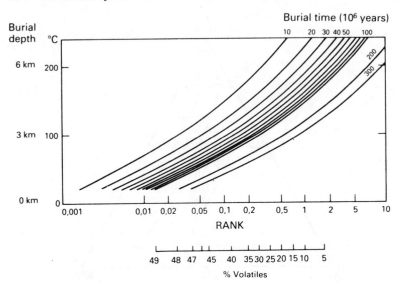

Fig. 5.14 Relationship between coal rank, temperature and time. The same rank can be produced by short burial times at high temperatures or long burial times at lower temperatures. Burial depth assumes average geothermal gradient of $35^{\circ}C\,km^{-1}$ (courtesy J. Karweil).

emphasises the importance of kinetics in determining the rank of coal produced. The situation is analogous to that in petroleum genesis where the most efficient generation of fossil fuels from an organic-rich sediment depends critically on the subsequent geological history of the deposit. Burial must position the sediment at the required depth, i.e. temperature, for the most favourable reaction and maintain that depth until complete reaction is achieved.

Glossary

The purpose of the following glossary is to provide a brief review of the main chemical principles used in the text. Each entry has been treated only in sufficient depth as to render the textbook self-explanatory, although this means that the level of treatment is somewhat arbitrary and uneven. It is hoped that the incorporation of a glossary will make the text accessible to students with a wider range of chemical ability, but it is recommended that one of the many excellent textbooks on physical and inorganic chemistry should be consulted for rigorous explanations.

The organisation of the glossary is such that the key words or phrases designated in the text have been grouped together under common headings.

Atomic Structure

Isotopes

An atom is the smallest particle of an element which can take part in a chemical change. Three main subatomic particles have been identified: the proton, neutron and electron. The proton has a mass nearly equal to that of the hydrogen atom and carries a unit positive charge. The neutron is an uncharged particle with a mass equal to that of the proton. The electron has a mass roughly 1/2000 of the mass of the proton but carries an equal amount of charge which is of opposite sign, i.e. negative.

Each atom contains a small positively-charged nucleus, made up of protons and neutrons, whilst the electrons are found in the outer regions of the atom where they comprise what amounts to a cloud of negative charge. All atoms contain an equal number (Z) of protons and electrons and this fundamental property of the element is known as its atomic number. Whereas the atomic number quantifies the presence of charged particles in an atom, the mass number determines the atomic weight.

Mass number = Number of protons + Number of neutrons

Clearly most of the atomic mass is present in the nucleus and it is possible to change the mass of an element, without changing the atomic number, by altering the number of neutrons. Since the chemical characteristics of an element are determined by Z, changes in the number of neutrons do not affect chemistry but only mass. Atoms of the same element which differ in the number of neutrons are called isotopes. Carbon, for example, possesses two stable isotopes which have mass numbers of 12 and 13, usually written as ^{12}C and ^{13}C. The isotope ^{12}C is much more abundant than ^{13}C, so that the average atomic weight of naturally occurring carbon is 12.011. A third isotope having mass number 14 (^{14}C) is unstable and undergoes radioactive decay.

Chemical Bonding

Simple theories of chemical bonding are based on the observation that the inert gases, helium (He), neon (Ne), argon (Ar), krypton (Kr) and xenon (Xe), are particularly unreactive. These elements are all monatomic gases, meaning that they exist in an unbonded state as single atoms, e.g. He or Ne. Furthermore, they show virtually no tendency to react with common laboratory reagents. The markedly unreactive behaviour suggests that their electronic structures are more stable than those of other elements. Theories of bonding assume that elements which do not possess these structures seek to attain them through their chemical reactions. Although there are different types of stable electronic structures, we can restrict ourselves to considering those elements which attempt to attain the inert gas structures. There are two main ways in which this occurs; by electron transfer (ionic bonding) and by electron sharing (covalent bonding).

Ionic bonding

Elements which possess electronic structures close to those of the inert gases may lose or gain the appropriate number of electrons to reach the inert gas structure. This mainly occurs in the following cases:

(1) Metals whose atoms have 1, 2 or 3 electrons more than an inert gas structure. Atoms of Group 1A and 2A metals acquire inert gas structures by losing one or two electrons respectively, so forming $1+$ and $2+$ ions (cations). The Group 3B metals and aluminium in Group 3A each have three more electrons than the most similar inert gas structure and form $3+$ ions.
(2) Non-metals whose atoms have 1 or 2 electrons less than the next inert gas structure. Non-metals in Group 6A gain two electrons to form $2-$ ions, those in Group 7A gain a single electron to form $1-$

ions. Nitrogen, the most electronegative element in Group 5A, forms 3 − ions sometimes. These negative ions are known as anions.

Therefore, an ionic bonded compound may be considered to be formed by transfer of electrons from a metallic element to a non-metallic element, with the oppositely charged ions held together by electrostatic attraction. For example, the element lithium ($Z = 3$) contains one more electron than the inert gas helium ($Z = 2$). Therefore, it forms the lithium ion Li^+ to attain the same electronic structure as helium:

$$Li \rightarrow Li^+ + e^-$$

By contrast, fluorine ($Z = 9$) contains one less electron than neon ($Z = 10$), so

$$F + e^- \rightarrow F^-$$

The compound lithium fluoride (LiF) is therefore formed by the transfer of an electron from lithium to fluorine, the resulting ions being held together in the LiF structure by electrostatic attraction. In the case of the metal calcium ($Z = 20$) two electrons must be lost to reach the nearest inert gas (argon, $Z = 18$) and, therefore, the calcium atom can supply electrons for two fluorine atoms and forms CaF_2. The relative proportions of cations and anions combined together is controlled by the requirement that all transferred electrons must be used and the final compound must be electrically neutral.

Covalent bonding

There are many chemical compounds whose existence cannot be satisfactorily explained by electron transfer. For example, chlorine gas exists as the molecule Cl_2 but there is no sensible way in which this structure could be obtained by ionic bonding. Remembering that elements seek to attain inert gas electronic structures through reaction, we can see that the chlorine atom ($Z = 17$) is one electron short of the structure of argon ($Z = 18$). The simplest explanation for the existence of Cl_2 is that each chlorine atom attains an inert gas structure by sharing one electron with its partner. One electron is contributed from each chlorine atom, so making a pair of shared electrons which constitute the covalent bond and which are counted towards the electronic structures of both atoms. By counting the pair of shared electrons twice, both atoms reach the inert gas configuration of argon.

The element chlorine is more typically found in ionic compounds, since it is a fairly simple matter for it to undergo reactions in which it gains a single electron to become the anion Cl^-. There are, however, many elements which would require the addition or removal of relatively large numbers of electrons in order to form ionic compounds. These elements characteristically form covalent bonds. Carbon ($Z = 6$) requires the addition of four electrons to reach the configuration of neon ($Z = 10$) or the removal of four electrons to reach the configuration of helium ($Z = 2$). Addition or removal of electrons on this scale is, however, energetically unfavourable, since work must be done to separate unlike charges or bring like charges together. The amount of energy required to remove each successive electron increases in each step:

$$C \xrightarrow[1.09\text{ kJ}]{-e^-} C^+ \xrightarrow[2.33\text{ kJ}]{-e^-} C^{2+} \xrightarrow[4.63\text{ kJ}]{-e^-} C^{3+} \xrightarrow[6.22\text{ kJ}]{-e^-} C^{4+}$$
$$Z = 6 \qquad Z = 5 \qquad Z = 4 \qquad Z = 3 \qquad Z = 2$$

The process of creating a C^{4-} ion from a carbon atom also requires the consumption of increasing amounts of energy. In contrast, electron sharing or covalency uses less energy to create an inert gas structure. The principle of covalent bonding is exactly the same however many pairs of electrons must be shared. Carbon requires a half share in four more electrons to reach an inert gas structure, which it achieves in CCl_4 by sharing a pair of electrons with each of the four chlorine atoms. Each chlorine atom is bonded to the carbon by donating a half share in one of its own electrons and receiving in exchange a half share in one of the carbon electrons. Thus the carbon atom has a half share in four of its own electrons plus a half share in one electron donated by each of the four chlorine atoms.

There are many other examples of covalent compounds, which include the numerous organic carbon compounds. In each case the relative proportions of elements so combined is determined by the requirement that each bond must represent a pair of shared electrons, contributed equally by the two atoms involved.

Electronegativity

There is a further complication in chemical bonding which has important environmental implications. In the example of chlorine gas, the two component atoms are identical and the pair of electrons constituting the covalent bond are shared equally. This is not always the case when the component atoms are dissimilar. For example, in HCl the chlorine atom has a strong affinity for electrons which are to a slight extent attracted away from the hydrogen and towards the chlorine. Both bonding electrons are still shared, but the sharing is unequal. Hence the chlorine atoms carry a slight negative charge and the hydrogen a corresponding positive charge. Such molecules in which the electrons are unsymmetrically distributed are called polar and the atom that carries the negative charge is said to be more electronegative. All degrees of polarity are possible from the zero polarisation seen in covalent compounds, where the atoms are both identical (e.g. Cl_2) and so have equal affinities for the bonding electrons, through intermediate polarities to the virtually complete polarisation found in ionic bonding. Therefore, there is no sharp distinction between covalent and ionic bonding, both are end members of a continuum which extends from equal sharing of electrons to complete transfer. This fact is emphasised by defining the percentage ionic character of a bond (Table 3.1). A bond is often called ionic if it has more than 50% ionic character, although it would be more correct to describe the bonding as predominantly ionic. Note that the reason for the dipolar nature of water is that oxygen is more electronegative than hydrogen and the bonding electrons are consequently displaced away from the hydrogen atoms and towards the oxygen (page 46).

Equilibrium

It is probable that all chemical reactions are reversible, i.e. can take place in both directions, although under earth surface conditions the extent of the reaction in one direction may predominate. Such reactions are then regarded as proceeding spontaneously to completion and the proportion of reactants converted into products is relatively large. However, the extent to which a reaction will proceed in any direction is influenced by concentration (dealt with in the section on Law of mass action, p. 160), temperature and pressure. A change in either or both temperature and pressure may reduce the extent of the forward reaction (reactants to products) and increase the backward reaction (products recombining to reactants), such that a mixture of both reactants and products are present after the overall reaction has ceased. For example, the reaction between two molecules of hydrogen and one molecule of oxygen to form water:

$$2H_2 + O_2 \xrightarrow{25°C} 2H_2O$$

An electric spark is necessary to surmount the activation energy barrier (p. 22) but the reaction is spontaneous under surface conditions and apparently results in complete conversion to water. However, at temperatures above 1500°C water vapour is decomposed to an appreciable extent into hydrogen and oxygen:

$$2H_2O \xrightarrow{1500°C} 2H_2 + O_2$$

Therefore, the backward reaction can be observed at high

temperatures and so reaction is demonstrably reversible, although the breakdown of water under surface conditions occurs to such a small degree that the reaction is normally said to proceed to completion.

At higher temperatures, where conditions are such that the forward and backward reactions can both occur to an appreciable extent, the reaction is best described as reversible. It has been found that after the lapse of a sufficient interval of time all reversible reactions reach a state of equilibrium, in which no further change in composition with time can be detected provided the conditions remain unaltered. The equilibrium is actually dynamic and the lack of change is due to the fact that the rates of both forward and backward reactions are equal.

Law of mass action

The law of mass action states that the rate of a chemical reaction is proportional to the molar concentrations of the reacting species, for solutes, and the partial pressure, for gases. For example, the reaction

$$H_2(g) + Br_2(aq) \rightleftharpoons 2HBr(aq)$$

According to the law of mass action, the rate of the forward reaction can be written as:

$$\text{Rate of forward reaction} \propto [H_2(g)][Br_2]$$
$$= k_1[H_2(g)][Br_2]$$

where the use of square brackets indicates concentration in molar units for solutes and $[H_2(g)]$ the partial pressure of hydrogen gas in atmospheres.

$$\text{Rate of backward reaction} \propto [HBr]^2$$
$$= k_2[HBr]^2$$

At equilibrium the rates of forward and backward reactions are equal:

$$k_1[H_2(g)][Br_2] = k_2[HBr]^2$$
$$K = \frac{k_1}{k_2} = \frac{[HBr]^2}{[H_2(g)][Br_2]}$$

The law of mass action, therefore, predicts that regardless of the initial concentrations of reactants, the position of the final equilibrium will be such that the ratio $[HBr]^2/[H_2(g)][Br_2]$ is constant.

The law of mass action can be expressed in more general terms by considering the reaction in which *a* moles of compound A react with *b* moles of compound B:

$$aA + bB \rightleftharpoons cC + dD$$

Then

$$K_E = \frac{[C]^c [D]^d}{[A]^a [B]^b}$$

The concentrations of each compound are expressed in molar units, e.g. mol 1^{-1}, for solutes, and partial pressures in atmospheres for gases. K_E is known as the equilibrium constant and varies with temperature and pressure. Values of K_E are usually given for standard conditions, i.e. 25° C or 298 K and 1 atm pressure.

Le Chatelier's principle

Le Chatelier's principle states that if a system at equilibrium is perturbed in any way, the system will react so as to minimise the effect of the perturbation. The following three chemical equilibria give examples of effects of changes in the concentrations of reactants or products, changes in volume and changes in temperature on the equilibrium position.

$$N_2 (g) + O_2 (g) \rightleftharpoons 2NO (g)$$

Le Chatelier's principle predicts that if reactant (N_2 or O_2) were added to, or product (NO) removed from, this equilibrium system then the forward reaction would occur to restore equilibrium by consuming reactant and creating more product. The converse would apply for removal of reactant or addition of more product.

The following reaction shows how changes in pressure can affect an equilibrium system.

$$N_2O_4 (g) \rightleftharpoons 2NO_2 (g)$$

When pressure is increased some of the NO_2 molecules combine with each other to form N_2O_4. Since two molecules of NO_2 are required to form one N_2O_4 molecule, the total number of molecules is reduced. This in turn reduces the pressure and partially compensates for the initial pressure increase. In general terms, changes in pressure act through a volume mechanism. Increasing pressure favours the production of smaller volumes of more dense materials.

In equilibrium systems where the temperature increases, Le Chatelier's principle predicts that the reaction which absorbs heat, and thus tends to minimise the temperature increase, will be favoured. Therefore, the addition of heat assists an endothermic reaction, in which heat is absorbed, whereas exothermic reactions, in which heat is evolved, are favoured by cooling.

Gas Solubility

Partial pressure and mole fraction

The relationship between the total pressure of a mixture of gases, e.g. air, and the pressures of the individual gases is expressed in the form of the law of partial pressures which states that the partial pressure of each gas in a mixture is defined as the pressure that gas would exert if it alone occupied the whole volume of the mixture at the same temperature. Therefore, the total pressure of a mixture of gases is equal to the sum of the partial pressures of the constituent gases. For each of the gases present in a mixture of volume V, therefore,

$$P_1 V = n_1 RT$$
$$P_2 V = n_2 RT$$
$$P_3 V = n_3 RT \text{ etc.}$$

where P_1 = partial pressure of gas 1, n_1 = number of moles of gas 1, R = gas constant, T = temperature, etc.
Hence

$$(P_1 + P_2 + P_3 \ldots)V = (n_1 + n_2 + n_3 \ldots)RT$$

But the sum of the partial pressures is equal to the total pressure P, therefore,

$$PV = (n_1 + n_2 + n_3 \ldots)RT = nRT$$

where $n = n_1 + n_2 + n_3 \ldots$

so

$$\frac{P_1 V}{PV} = \frac{n_1 RT}{nRT}$$

or

$$P_1 = \frac{n_1}{n} P$$

The fraction $\left(\dfrac{n_1}{n} \right)$ is the mole fraction of constituent 1 in the mixture. The mole fraction is defined as the number of moles of that constituent divided by the total number of moles in the mixture.

Thus, the partial pressure of a gas is derived from its mole fraction and the total pressure. The mole fractions of N_2, O_2 and Ar in air are 0.78, 0.21 and 0.01 respectively, and these are also their partial pressures at a total pressure of 1 atm.

Henry's law

The influence of pressure on the solubility of a gas is expressed by Henry's law which states that the mass of gas dissolved by a given volume of solvent, at constant temperature, is proportional to the pressure of the gas in equilibrium with the solution. Provided the pressures are not too high or the temperatures are not too low, most gases obey Henry's law, especially if they are not very soluble. When chemical reaction takes place between the gas and solvent, e.g. CO_2 in water, Henry's law fails to hold if the total solubility is considered. However, the amount of gas in a free or unreacted state does increase in proportion to the pressure as predicted by Henry's law. When gases dissolve from a mixture, instead of from the pure gas, the solubility of each component is proportional to its own partial pressure. Therefore, Henry's law applies to each gas independently of the pressure of the other gases present in the mixture. The constant of proportionality relating solubility to partial pressure is known as the Henry's law constant, K_H. For the solution of nitrogen in water:

$$N_2(g) \rightleftharpoons N_2(aq)$$

$$K_H = \frac{[N_2(aq)]}{P_{N_2}} = \frac{[N_2(aq)]}{[N_2(g)]}$$

Since partial pressure and mole fraction are related, the Henry's law constant K_H is the same as the equilibrium constant K_E obtained by writing the law of mass action for the solution of nitrogen in water.

Geothermal Gradient

The increase in temperature with depth from the surface of the earth, caused by heat flow from its interior, is called the geothermal gradient. The decay of radioactive elements in the interior produces a heat flow of 8×10^{20} J yr^{-1} to the surface. However, the heat felt at the surface comes mainly from the sun because the thermal conductivity of the crust is so low the flow of heat from the interior occurs only slowly. The low thermal conductivity not only prevents heat from the interior escaping, but also effectively prevents solar heat from penetrating into the crust. Below 30 m depth seasonal variations in solar heat have little influence on crustal temperatures and the average geothermal gradient is approximately 35 K km^{-1}.

Hydrogen Bonds

Hydrogen exhibits an unusual type of bonding in certain circumstances. Although the hydrogen atom possesses only one electron, and is therefore normally univalent, it is capable of forming a bond or bridge between two atoms which are small and strongly electronegative e.g. F, O, N. The most common hydrogen bond is formed between hydroxyl groups on adjacent molecules. Each hydroxyl bond is polarised, due to the electronegative character of the oxygen atom, so the bonding electrons are markedly displaced from the hydrogen atom. Consequently these hydrogen atoms behave almost like protons and can form intermolecular bonds with the oxygen atoms on adjacent hydroxyl groups. Each water molecule can be surrounded by four others, connected to the central one by hydrogen bonds, because each oxygen can bond to two hydrogen atoms in different water molecules and the two hydrogens can each bond to oxygen in different water molecules. This structure actually exists in ice (Fig. G.1).

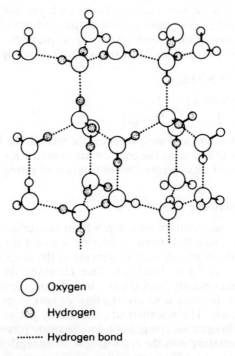

○ Oxygen

⊙ Hydrogen

········ Hydrogen bond

Fig. G.1 The structure of ice (from Coxon, J. M., Fergusson. J. E. and Phillips. L. F., (1980) *First Year Chemistry*, Edward Arnold, London).

Isomorphous Substitution

The term isomorphism is applied to the phenomena of substances with analogous formulae having closely related crystal structures. For example, $CaCO_3$ (calcite) and $MnCO_3$ (rhodocrosite) are isomorphous and the similarity in structure allows a certain degree of interchangeability between the two cations in these compounds. Thus some calcium ions in calcite may be replaced by manganese ions and vice versa. The substitution of some Ca^{2+} by Mn^{2+} may be indicated in the mineral formula by writing $(Ca, Mn) CO_3$. Because minerals often crystallise from complex fluids containing a wide variety of ions, the crystal lattice may have the opportunity to accept many trace elements into it by isomorphous substitution. Hence minerals rarely correspond to their ideal chemical formulae.

Isomorphism is caused by the fact that cations and anions of the same relative size, present in the same proportions in a compound, tend to surround themselves with the same number of nearest neighbours, as determined by the radius ratio rule (p. 50). Therefore, they crystallise into similar types of structure. The most important factor in isomorphism is the similarity in size relations of the different ions rather than any chemical similarity. The Mg^{2+} and Fe^{2+} ions have roughly the same ionic radius ($0.66Å$ and $0.74Å$, respectively) so that an isomorphous relationship between magnesium and iron is possible in many silicates, e.g. olivines, pyroxenes and amphiboles. It does not follow that all pairs of magnesium and iron-bearing compounds are isomorphous, e.g. very little magnesium is found in pyrite (FeS_2) and very little iron in epsomite ($MgSO_4.7H_2O$). The lack of isomorphism in these compounds is attributable to the more electronegative behaviour of iron, in forming bonds with a greater degree of covalent character than magnesium. With increasing covalency, simple packing characteristics, as predicted by the radius ratio rule, are modified by spatial requirements imposed by the need to share electrons between atoms.

Organic Compounds

Organic compounds are classified into three major groups:

(1a) *Aliphatic* compounds, which have an open chain molecular structure, e.g. the paraffins (Fig. 5.9).
(b) *Alicyclic* compounds, which contain closed rings of carbon atoms only, e.g. the cycloparaffins (Fig. 5.9), and resemble aliphatic compounds in many of their properties.
(2) *Aromatic* compounds, which are ring compounds containing at

$$
\begin{array}{c}
\text{H} \\
\text{C} \\
/ \quad \backslash\!\backslash
\end{array}
$$

least one benzene ring
$$
\begin{array}{c}
\text{HC} \qquad \text{CH} \\
\| \qquad\quad | \\
\text{HC} \qquad \text{CH} \\
\backslash \qquad /\!/ \\
\text{C} \\
\text{H}
\end{array}
$$
which can be represented as

The structures of some aromatic compounds are given in Fig. 5.9.
(3) *Heterocyclic* compounds, which are ring compounds containing other elements such as oxygen, sulphur and nitrogen (Fig. 5.9) besides carbon in the ring.

Phases

A phase is defined as any homogeneous and physically distinct part of a system which is separated from other parts of the system by definite bounding surfaces. For example, ice, liquid water and water vapour are three phases of water since each is physically distinct and homogeneous and there are definite boundaries between them. The phases participating in chemical reactions in the text are identified by the letters (s), (l) and (g) for solid, liquid and gas respectively. Only reactions in which there is a phase change are so denoted.

Redox Reactions: Oxidation, Reduction, Oxidation State, Oxidation Number, Redox Potential

Many chemical reactions involve the transfer of electrons from one species to another. A simple example is the formation of Li^+ and F^- ions from their corresponding atoms to form the compound LiF:

$$
\begin{array}{l}
Li \rightarrow Li^+ + e^- \\
F + e^- \rightarrow F^-
\end{array}
$$

The lithium atom has transferred an electron to the fluorine atom and any reaction between atoms which leads to the formation of ions may be similarly explained. The overall reaction can be derived by summing the two components which are each known as a half-reaction:

Half-reaction	$Li \rightarrow Li^+ + e^-$	An electron appears on each
Half-reaction	$F + e^- \rightarrow F^-$	side of the overall equation
Overall Reaction	$Li + F \rightarrow Li^+ + F^-$	and can be cancelled.

The half-reaction in which the lithium atom loses an electron is referred to as an *oxidation* half-reaction and the half-reaction in which the fluorine atom accepts an electron (the one donated by lithium) is known as a *reduction* half-reaction. Lithium is then said to be oxidised and flourine to be reduced. Further, since electrons must be conserved, i.e. transferred without loss from one substance to another, it is clear that oxidation and reduction must always occur together. Hence the overall reaction is known as an oxidation-reduction or redox reaction.

Some elements, for example iron, may undergo a sequence of oxidation steps. Iron, existing as the pure element Fe^0, may be oxidised first to the Fe^{2+} ion and then to the Fe^{3+} ion.

$$Fe \rightarrow \quad Fe^{2+} + 2e^-$$

Pure   Iron (II)

element

$$Fe^{2+} \rightarrow Fe^{3+} + e^-$$

Iron (III)

These different forms in which iron may exist represent different degrees of oxidation, which are known as *oxidation states*. Each different oxidation state is distinguished by its *oxidation number*. For example, the oxidation number of ferrous iron is $+2$, whether the ferrous iron occurs in aqueous solution as the free Fe^{2+} ion, or combined in the compounds $FeCl_2$, FeO etc. In the case of a covalent compound the oxidation number refers to the charge which that atom would have if all the bonding electrons were arbitrarily assigned to the more electronegative atom. Therefore, the oxidation number of iron in FeS is $+2$. A similar point can be made by distinguishing the terms Fe(II) and Fe^{2+}. The former refers to the oxidation number of iron and can be applied to the free ion and both ionic and covalent compounds. The latter is only used in situations where the Fe^{2+} ion is actually present, that is as the free ion or in an ionic-bonded solid. It is incorrect to refer to the form of iron in FeS as Fe^{2+}.

In the section on the oxygen cycle (p. 13) a redox equation is presented for the oxidation of Fe^{2+} to $Fe(OH)_3$:

$$2Fe^{2+} + \tfrac{1}{2}O_2 + 5H_2O \rightarrow 2Fe(OH)_3 + 4H^+$$

Fe(II)   Fe(III)

Inspection of this equation readily shows the oxidation of Fe(II) to Fe(III) and the electrons thus generated are used to reduce oxygen:

$$\tfrac{1}{2}O_2 + H^+ + 2e^- \rightarrow OH^-$$

The hydrogen ion required in this reaction and the remaining hydroxyl ions needed for the formation of $Fe(OH)_3$ are generated by the dissociation of water:

$$H_2O \rightarrow H^+ + OH^-$$

In discussing redox reactions the phrases oxidising agent and reducing agent are frequently used to refer to the species responsible for oxidation and reduction. In the iron case above oxygen is spoken of as an oxidising agent, since it causes the oxidation of Fe(II) to Fe(III). By the same token, Fe(II) is a reducing agent and is responsible for the reduction of atmospheric oxygen from an oxidation number of 0 to -2. This point can be neatly summarised by stating that, in a redox reaction, oxidising agents are reduced and reducing agents are oxidised.

In many environmental redox reactions it is difficult to decide precisely which atoms are losing or gaining electrons. The concept of oxidation number is helpful here. Oxidation is defined as an increase in oxidation number and reduction as a decrease in oxidation number. Assigning oxidation numbers to atoms in chemical species can be done using certain arbitrary rules.

(1) The oxidation number of all elements is 0.
(2) The oxidation number of a monatomic ion is equal to the charge on that ion, e.g. $Na^+ = Na(+1)$, $Al^{3+} = Al(+3)$, $Cl^- = Cl(-1)$.
(3) Oxygen has an oxidation number of -2 in all compounds except O_2, peroxides and superoxides.
(4) Hydrogen has an oxidation number of $+1$ in all compounds.
(5) The sum of the oxidation numbers of the elements in a compound or ion equals the charge on that species.
(6) The oxidation number of the elements in a covalent compound can be deduced by considering the shared electrons to belong exclusively to the more electronegative atom. Where both atoms have the same electronegativity the electrons are considered to be equally shared. Thus the oxidation numbers of carbon and chloride in CCl_4 are $+4$ and -1 respectively, and the oxidation number of chlorine in Cl_2 is 0.

The stability of an element in a particular oxidation state depends on the energy change involved in adding or removing electrons. A quantitative measure of this energy change is provided by the electrode potential. The electrode potential of any reaction is a relative figure, the reference being the reaction

$$2H^+ + 2e^- \rightarrow H_2$$

Under standard conditions the electrode potential of this reaction is assigned the value 0.00 volts. Reactions which involve a change in the oxidation number of an element are usually written with the more oxidised form on the left, e.g.

$$Zn^{2+} + 2e^- \rightarrow Zn$$

These reactions can be listed in order of increasing electrode potential, under standard conditions, (Table G.1) and their relative positions determine the ease with which oxidation or reduction may occur. The reduced form of a listed reaction will react with, i.e. will reduce, the oxidised form of any reaction below it, but not with the oxidised form of a reaction above it.

In any natural system there may be a wide variety of species present, each of which may possess the capacity to act as an oxidising or reducing agent to a different degree. The ability of the overall system to act as an oxidising or reducing agent is known as its *redox potential* , and is determined by the electrode potentials between the oxidised and reduced species present and their concentrations. In effect the redox potential measures the ability of the environment to supply electrons to an oxidising agent or take up electrons from a reducing agent.

Thermodynamics

Thermodynamics literally means the dynamics of thermal changes. It deals with relations between heat and other forms of energy and is used to describe the physical and chemical changes which matter can undergo. That part of thermodynamics which describes chemical changes, which is the main concern here, is known as thermochemistry. The basis for thermochemical calculations arises out of the more general statements of thermodynamics, as expressed in the First and Second Laws of Thermodynamics.

Free Energy and the first law of thermodynamics

The first law of thermodynamics states that energy may be transformed from one form to another, but it cannot be destroyed, i.e. the total energy of a system must remain constant, although the energy in the system may be present in different forms. In each chemical reaction a certain amount of energy is associated with the structure of the reactants and a certain amount with the products and the difference between these two values represents the amount of energy used or liberated in the reaction. This difference is called the change in *free energy* (ΔG°). The superscript indicates that the ΔG value has been measured under standard conditions (unit activity of dissolved species, 298 K and 1 atmosphere pressure). The change in free energy can be measured for a large number of reactions. In the example below the energy liberated appears as 131 100 Joules of heat:

$$\tfrac{1}{2}H_2 + \tfrac{1}{2}Cl_2 \rightarrow HCl \qquad \Delta G^{\circ} = -131.1 \text{ kJ}$$

By arbitrarily assigning the value zero to all pure elements at

Table G.1 Some standard electrode potentials (adapted from Coxon, J. M., Fergusson, J. E. and Phillips, L. F., (1980) *First Year Chemistry*, Edward Arnold, London)

Reaction		Electrode potential (volts)
$Li^+ + e^-$	$\rightarrow Li(s)$	-3.04
$K^+ + e^-$	$\rightarrow K(s)$	-2.92
$Ca^{2+} + 2e^-$	$\rightarrow Ca(s)$	-2.87
$Na^+ + e^-$	$\rightarrow Na(s)$	-2.71
$Mg^{2+} + 2e^-$	$\rightarrow Mg(s)$	-2.36
$Al^{3+} + 3e^-$	$\rightarrow Al(s)$	-1.66
$Mn^{2+} + 2e^-$	$\rightarrow Mn(s)$	-1.18
$Zn^{2+} + 2e^-$	$\rightarrow Zn(s)$	-0.76
$Cr^{3+} + 3e^-$	$\rightarrow Cr(s)$	-0.74
$S(s) + 2e^-$	$\rightarrow S^{2-}$	-0.48
$Fe^{2+} + 2e^-$	$\rightarrow Fe(s)$	-0.41
$Co^{2+} + 2e^-$	$\rightarrow Co(s)$	-0.28
$Ni^{2+} + 2e^-$	$\rightarrow Ni(s)$	-0.23
$Sn^{2+} + 2e^-$	$\rightarrow Sn(s)$	-0.14
$Pb^{2+} + 2e^-$	$\rightarrow Pb(s)$	-0.13
$Fe^{3+} + 3e^-$	$\rightarrow Fe(s)$	-0.02
$H^+ + e^-$	$\rightarrow \frac{1}{2}H_2(g)$	0.00
$Cu^{2+} + 2e^-$	$\rightarrow Cu(s)$	$+0.35$
$Cu^+ + e^-$	$\rightarrow Cu(s)$	$+0.52$
$\frac{1}{2}I_2(s) + e^-$	$\rightarrow I^-$	$+0.54$
$Ag^+ + e^-$	$\rightarrow Ag(s)$	$+0.80$
$Hg^{2+} + 2e^-$	$\rightarrow Hg(l)$	$+0.85$
$\frac{1}{2}Br_2(l) + e^-$	$\rightarrow Br^-$	$+1.07$
$MnO_2(s) + 4H^+ + 3e^-$	$\rightarrow Mn^{2+} + 2H_2O$	$+1.23$
$\frac{1}{2}Cr_2O_7^{2-} + 7H^+ + 3e^-$	$\rightarrow Cr^{3+} + 3\frac{1}{2}H_2O$	$+1.33$
$\frac{1}{2}Cl_2(g) + e^-$	$\rightarrow Cl^-$	$+1.36$
$PbO_2(s) + 4H^+ + 2e^-$	$\rightarrow Pb^{2+} + 2H_2O$	$+1.46$
$MnO_4^- + 8H^+ + 5e^-$	$\rightarrow Mn^{2+} + 4H_2O$	$+1.51$
$MnO_2^- + 4H^+ + 3e^-$	$\rightarrow MnO_2(s) + 2H_2O$	$+1.70$
$\frac{1}{2}F_2(g) + e^-$	$\rightarrow F^-$	$+2.87$

conditions of 298 K and 1 atmosphere, the free energy of formation HCl is determined as -131.1 kJ mol^{-1}. Similar means can be used to measure the free energy of formation for a wide range of compounds and the resulting lists of $\Delta G^{\ominus}$ values can be used to determine the free energy changes in an almost infinite number of reactions. Note that because all pure elements have been given the same arbitrary value of zero, this approach gives no guidance as to the absolute free energy of a system, it only predicts changes in free energy. This, however, has not limited the usefulness of free energy values. It is a consequence of the first law of thermodynamics that the energy changes in a reaction are independent of the number of steps by which that reaction is carried out. Therefore, values initially derived from reactions involving pure elements can be correctly used to predict the energy changes in reactions involving compounds of these elements (see p. 167). This allows listed values of free energies of formation to be used to predict the course of chemical reactions. It is found that those reactions which liberate energy ($\Delta G^\ominus$ negative) take place spontaneously, whereas reactions which have a positive value of $\Delta G^\ominus$ need an external supply of energy before the products can be formed. For values of $\Delta G^\ominus = 0$ the reaction is at equilibrium.

One of the most useful applications of free energy changes is to derive values for the equilibrium constant of a reaction using the following relationship:

$$\log_{10} K = \frac{-\Delta G^\ominus}{2.303\ RT}$$

where R = Gas constant = 8.314 Joules degree^{-1} mole^{-1}
T = Temperature in degrees Kelvin
At 25° C or 298 K,

$$\log_{10} K_E = \frac{-\Delta G^\ominus}{5.707}$$

The value for K_E so derived is for a reaction carried out at 298 K. If K_E is required at other temperatures a different value for T must be used in the first relationship and the free energy change ΔG must also be recalculated at the appropriate temperature, since $\Delta G^\ominus$ values are derived under standard conditions.

The second law of thermodynamics

The second law of thermodynamics states that any closed system becomes increasingly disordered with time. Disorder describes the extent to which the particles of a substance are confined to a given region of space. The greater the degree of disorder, the less the particles are confined. In a closed system natural processes tend to eradicate the physical and chemical differences between regions, on both microscopic and macroscopic scales. These processes are spontaneous and can only be reversed with an energy input, that is if the system is no longer isolated.

The different forms of energy used in the earth-air-water factory are solar radiation, mechanical energy, chemical energy and the earth's heat content. Solar radiation is the only source external to the earth and is responsible for creating the chemical and physical heterogeneity around us. For example, solar radiation creates the temperature differences, wind systems and rainfall patterns, etc., which develop climatic regimes. Solar radiation is also converted by photosynthesis into the chemical energy stored in coal, oil and living matter, which can later be used in other chemical reactions. In the absence of solar radiation climatic differences would even out and the creation of stored chemical energy would cease. Therefore, the characteristic physical and chemical heterogeneity of the earth is created by solar input, and without this heterogeneity disorder would prevail.

Bibliography

A section of recommended elementary texts precedes the main references which are listed by chapter and in alphabetical order. General reading on geochemistry is listed with Chapter 1 and a supplementary section of more advanced texts is given at the end.

Elementary Chemistry Texts

COXON, J. M., FERGUSSON, J. E. and PHILIPS, L., (1980) *First Year Chemistry*, Edward Arnold, London.

GARVIE, D., REID, J. and ROBERTSON, A., (1976) *Core Chemistry*, Oxford University Press, Oxford.

MASTERTON, W. L. and SLOWINSKI, E. J., (1977) *Chemical Principles*, W. B. Saunders, Eastbourne.

MORRIS, J. G., (1974) *A Biologist's Physical Chemistry*, Edward Arnold, London.

Chapter 1 Introduction

FYFE, W. S., (1974) *Geochemistry*, Clarendon Press, Oxford.

GARRELS, R. M. and MACKENZIE, F. T., (1971) *The Evolution of Sedimentary Rocks*, Norton, New York.

KRAUSKOPF, K. B., (1967) *Introduction to Geochemistry*, McGraw Hill, New York.

SIEVER, R., (1974) The steady state of the earth's crust, atmosphere and oceans. *Scientific American*, **230**, 72–79.

TUREKIAN, K. K., (1972) *Chemistry of the Earth*, Holt, Rinehart and Winston, New York.

Chapter 2 The Atmosphere

BROECKER, W. S., (1970) Man's oxygen reserves. *Science*, **168**, 1537–1538.

GARRELS, R. M., LERMAN, A. and MACKENZIE, F. T., (1976) Controls of atmospheric O_2 and CO_2: past, present and future. *American Scientist*, **64**, 306–315.

HAGGEN-SMIT, A. J., (1964) The control of air pollution. *Scientific American*, **210**, 25–31.

HAMILTON. C. L.. (1973) *Chemistry in the Environment*, Readings from Scientific American, Freeman and Co., San Francisco.

JUNGE. C. E.. (1974) Residence time and the variability of tropospheric trace gases. *Tellus*, **26**, 477–488.

JUNGE. C. E. and WERBY. R. T.. (1958) The concentration of chloride, sodium, potassium, calcium and sulphate in rainwater over the United States. *Journal of Meteorology*, **15**, 417–425.

LIU. S. C.. CICERONE. R. J. and DONAHUE. T. M.. (1977) Sources and sinks of atmospheric N_2O and the possible ozone reduction due to industrial fixed nitrogen fertilizers. *Tellus*, **29**, 251–263.

MACHTA. L.. (1972) The role of the oceans and biosphere in the carbon dioxide cycle. In: *The Changing Chemistry of the Oceans*, Ed. D. Dryssen and D. Jayner, Wiley Interscience, London.

SAWYER. J. C.. (1972) Man-made carbon dioxide and the 'greenhouse' effect. *Nature*, **239**, 23–26.

SPEDDING. D. J.. (1974) *Air Pollution*, Clarendon Press, Oxford.

WILKINS. E. T.. (1954) Air pollution aspects of the London fog of December 1952. *Quarterly Journal of the Royal Meteorological Society*, **80**, 267–271.

WOODWELL. G. M.. (1978) The carbon dioxide question. *Scientific American*, **238**, 38–43.

Chapter 3 The Crust

CROMPTON. E.. (1962) Soil Formation. *Outlook on Agriculture*, **3**, 209–218.

CURTIS. C. D.. (1970) Differences between lateritic and podzolic weathering. *Geochimica et Cosmochimica Acta*, **34**, 1351–1353.

CURTIS. C. D.. (1975) Chemistry of rock weathering: fundamental reactions and controls. In: *Geomorphology and Climate*, Ed. E. Derbyshire, Wiley, London.

FITZPATRICK. E. A.. (1971) *Pedology: A Systematic Approach to Soil Science*, Chapters 1, 2, 3, 4 and 8, Oliver and Boyd, Edinburgh.

GIBBS. R. J.. (1970) Mechanisms controlling world water chemistry. *Science*, **170**, 1088–1090.

MACAN. T. T.. (1970) *Biological Studies of the English Lakes*, Chapters 4 and 5, Longmans, London.

OLLIER. C. D.. (1962) *Weathering*, Oliver and Boyd, Edinburgh.

SHERMAN. G. D.. (1952) The genesis and morphology of alumina-rich laterite clays. In: *Problems in Clay and Laterite Genesis*, American Institute of Mining, Metallurgical and Petroleum Engineers, p. 154.

Chapter 4 The Oceans

BERNER, R. A., (1971) *Principles of Chemical Sedimentology*, McGraw Hill, New York.

BIDLEMAN, T. E., RICE, C. P. and OLNEY, C. E., (1976) High molecular weight chlorinated hydrocarbons in the air and sea: rates and mechanism of air/sea transfer. In: *Marine Pollution Transfer*, Ed. H. L. Windom and R. A. Duce, Lexington Books, Lexington, pp. 323–351.

BROECKER, W. S. and TAKAHASHI, T., (1966) Calcium carbonate precipitation on the Bahama Banks. *Journal of Geophysical Research*, **71**, 1575–1602.

GOLDBERG, E. D., (1976) *The Health of the Oceans*, Unesco Press, Paris.

HEATH, G. R., (1974) Dissolved silica and deep sea sediments. In: *Studies in Paleo-Oceanography*, Ed. W. W. Hay, Society of Economic Paleontologists and Mineralogists Special Publication No. 20, Tulsa, pp. 77–93.

KINSMAN, D. J. J., (1975) Salt floors to geosynclines. *Nature*, **255**, 375–378.

LOVELOCK, J. E., MAGGS, R. J. and WADE, R. J., (1973) Halogenated hydrocarbons in and over the Atlantic. *Nature*, **241**, 194–196.

MACINTYRE, F., (1970) Why the sea is salt. *Scientific American*, **223**, 104–115.

MACKENZIE, F. T. and GARRELS, R. M., (1966) Chemical mass balance between rivers and oceans. *American Journal of Science*, **264**, 507–525.

THRUSH, B. A., (1977) The chemistry of the stratosphere and its pollution. *Endeavour*, **1**, 3–6.

Chapter 5 Formation of Mineral Resources in Sediments

ALLEN, J. R. L., (1964) The Nigerian continental margin: bottom sediments, submarine morphology and geological evolution. *Marine Geology*, **1**, 289–332.

BERNER, R. A., (1971) *Principles of Chemical Sedimentology*, McGraw Hill, New York.

BROECKER, W. S., (1974) *Chemical Oceanography*, Harcourt, Brace, Jovanovich, New York.

CLAYPOOL, G. E. and KAPLAN, I. R., (1974) The origin and distribution of methane in marine sediments. In: *Natural Gases in Marine Sediments*, Ed. I. R. Kaplan, Plenum Press, New York.

CURTIS, C. D., (1977) Sedimentary geochemistry; environments and processes dominated by an aqueous phase. *Philosophical Transactions of the Royal Society (London)*, **286A**, 353–372.

176   *Bibliography*

FORSMAN, J. P. and HUNT, J. M., (1959) Insoluble organic matter (kerogen) in sedimentary rocks. *Geochimica et Cosmochimica Acta*, **15**, 170–182.

RILEY, J. P. and CHESTER, R., (1971) *Introduction to Marine Chemistry*, Academic Press, London.

For More Advanced Students

BERNER, R. A., (1971) *Principles of Chemical Sedimentology*, McGraw Hill, New York.

CAMPBELL, I. M., (1977) *Energy and the Atmosphere*, Wiley, London.

GARRELS, R. M., MACKENZIE, F. T. and HUNT, C., (1975) *Chemical Cycles and the Global Environment*, Kaufmann, California.

RILEY, J. P. and CHESTER, R., (1971) *Introduction to Marine Chemistry*, Academic Press, London.

STUMM, W. and MORGAN, J. J., (1970) *Aquatic Chemistry*, Wiley Interscience, London.

Index